Transformer and Inductor
Design Handbook

Additional Volumes in Preparation

Electrical Engineering-Electronics Software

1. Transformer and Inductor Design Software for the IBM PC,
   Colonel Wm. T. McLyman
2. Transformer and Inductor Design Software for the Macintosh,
   Colonel Wm. T. McLyman
3. Digital Filter Design Software for the IBM PC,
   Fred J. Taylor and Thanos Stouraitis

Transformer and Inductor Design Handbook

Second Edition, Revised and Expanded

Colonel Wm. T. McLyman

Jet Propulsion Laboratory
California Institute of Technology
Pasadena, California

Marcel Dekker, Inc. New York and Basel

Library of Congress Cataloging-in-Publication Data

McLyman, Colonel William T.
  Transformer and inductor design handbook.

  (Electrical engineering and electronics ; 49)
  Includes bibliographies and index.
    1. Electronic transformers—Design and construction
—Handbooks, manuals, etc. 2. Electric inductors—
Design and construction—Handbooks, manuals, etc.
I. Title. II. Series
TK7872.T7M32 1988   621.3815'3   88-3728
ISBN 0-8247-7828-6

MARCEL DEKKER, INC.
270 Madison Avenue, New York, New York 10016

Current printing (last digit):
10 9 8 7 6 5 4 3 2 1

PRINTED IN THE UNITED STATES OF AMERICA

To my wife
Olga

Preface

About seven years ago, I started giving six seminars per year around the country, speaking on modern methods to design magnetic components, at which I introduced new equations and showed how they minimize the design time. The participants at these seminars requested that I write the second edition of the *Transformer and Inductor Design Handbook*, including the new equations and their relationships to the design of transformers and inductors along with many design examples.

This book offers a practical approach for design engineers and systems engineers in the electronics industry, as well as the aerospace industry. Tranformers are to be found in virtually all electronic circuits. This book can easily be used to design lightweight, high-frequency aerospace tranformers or low-frequency commercial transformers. It is, therefore, a design manual.

The conversion process in power electronics requires the use of transformers, components which frequently are the heaviest and bulkiest item in the conversion circuit. They also have a significant effect on the overall performance and efficiency of the system. Accordingly, the design of such transformers has an important influence on overall system weight, power conversion efficiency, and cost. Because of the interdependence and interaction of parameters, judicious tradeoffs are necessary to achieve design optimization.

Manufacturers have for years assigned numeric codes to their cores to indicate their power-handling ability. This method assigns to each core a number called the area product A_p that is the product of its window area W_a and core cross-section area A_c. These numbers are used by core suppliers to summarize dimensional and electrical properties in their catalogs. The product of the window area W_a and the core area A_c, gives the area product A_p a dimension to the fourth power. I have developed a new equation for the power-handling ability of the core, the core geometry K_g. The core goemetry K_g has a dimension to the fifth power. This new equation gives engineers faster and tighter control of their design. The core geometry coefficient K_g is a relatively new concept, and magnetic core manufacturers are now beginning to put it in their catalogs.

Because of their significance, the area product A_p and the core geometry K_g are treated extensively in this handbook. A great deal of other information is also presented for the convenience of the designer. Much of the material is in tabular form to assist the designer in making the tradeoffs best suited for the particular application in a minimum amount of time.

Designers have used various approaches in arriving at suitable transformer and inductor designs. For example, in many cases a rule of thumb used for dealing with current density is that a good working level is 1000 circular mils per ampere. This is satisfactory in many instances; however, the wire size used to meet this requirement may produce a heavier and bulkier inductor than desired or required. The information presented here will make it possible to avoid the use of this and other rules of thumb and to develop a more economical and better design. While there are other books available on electronic transformers, none seems to have been written with the user's viewpoint in mind. The material is organized so that the student engineer or technician, starting at the beginning of the book and continuing through to the end, will gain a comprehensive knowledge of the state of the art in transformer and inductor design.

No responsibility is assumed by the author or the publisher for any infringement of patent or other rights of third parties which may result from the use of circuits, systems, or processes described or referred to in this handbook.

I wish to thank the manufacturers represented in this book for their assistance in supplying technical data.

Colonel Wm. T. McLyman

Contents

Chapter 1

Fundamentals of Magnetics and the Selection of Materials for Static Inverter and Converter Transformers

1.1 FUNDAMENTALS OF MAGNETICS

Considerable difficulty is encountered in mastering the field of magnetics because of the use of so many different systems of units—the centimeter-gram-second (cgs) system, the meter-kilogram-second (mks) system, and the mixed English units system. Magnetics can be treated in a simple way by using the cgs system.

1.1.1 Magnetic Properties in Free Space

A long wire with a dc current I flowing through it produces a circulatory magnetic field H around the conductor as shown in Figure 1.1, where relationship is

$$H = \frac{Ni}{\ell} = \frac{i}{\ell}$$

cgs

$$H = \frac{AMPERES}{cm}$$

Note per Electomagn. J. Kraus

$H =$ amperes/m
$= A\,m^{-1}$
Above is SI units.

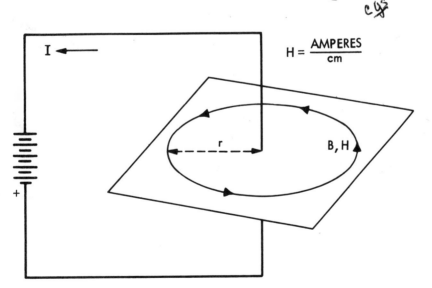

Figure 1.1. Magnetic field.

right hand rule is applied here for direction of magnetic field

1

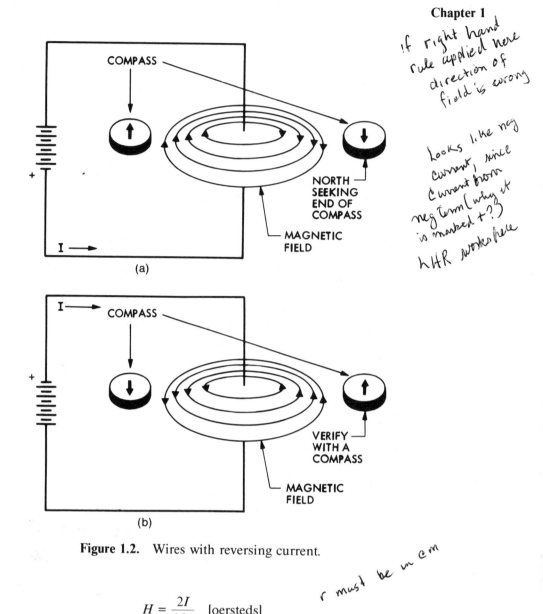

If right hand rule applied here direction of field is wrong

Looks like neg current, since current from neg term (why it is marked +?) LHR works here

Figure 1.2. Wires with reversing current.

$$H = \frac{2I}{10r} \quad \text{[oersteds]}$$

r must be in cm

When a current is passed through the wire in one direction as shown in Figure 1.2(a), the needle in the compass will point in one direction. When the current in the wire is reversed, Figure 1.2(b), the needle will also reverse direction. This shows that the magnetic field has polarity and that when the current I is reversed the magnetic field H will follow the current reversals.

If the wire is wound on a dowel, its magnetic field is greatly intensified. The coil, in fact, exhibits a magnetic field exactly like that of a bar magnet shown in Figure 1.3. Like the bar magnetic, the coil has a north pole and a neutral center region. Moreover, the polarity can be reversed by reversing the current I through the coil. This again demonstrates the dependence of the direction of the magnetic field on the current direction.

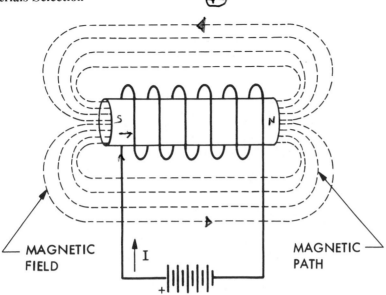

Figure 1.3. Air coil with an intensified magnetic field.

The magnetic circuit is the space in which the flux flows around the coil. The magnitude of the flux is determined by the product of the current and the number of turns N in the coil. The force NI required to create the flow is called the *magnetomotive force* (mmf). Figure 1.4 shows the relationship between flux density B and magnetizing force H for an air-core coil. The ratio of B to H is called the permeability μ, and for this air-core coil the ratio is unity in the cgs system, where it is expressed in units of gauss per oersted (G/Oe). $\mu = 1\ G/Oe$

If the battery in Figure 1.3 were replaced with an ac source as shown in Figure 1.5, the relationship between B and H would have the characteristic shown in Figure 1.6. The linearity of the relationship between B and H represents the main advantage of air-core coils. Since the relationship is linear, increasing H increases B and therefore the flux in the coil, and in this way very large fields can be produced with large currents. There is obviously a practical limit to this, which depends on the maximum allowable current in the conductor and the resulting temperature

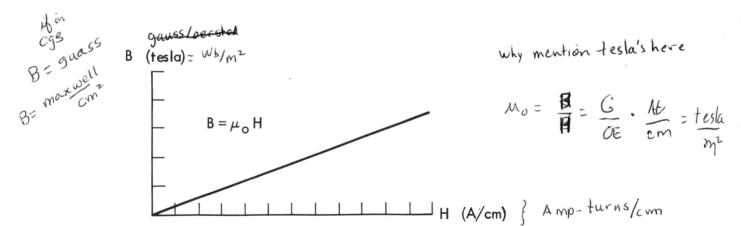

Figure 1.4. Relationship between B and H.

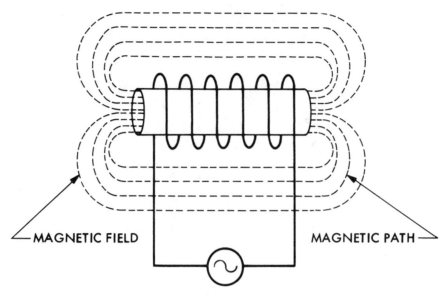

Figure 1.5. Air coil drive from an ac source.

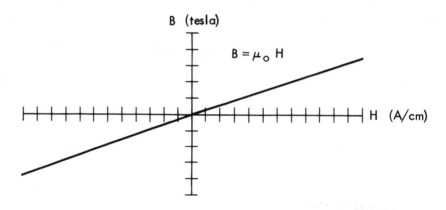

Figure 1.6. Relationship between B and H with ac excitation.

rise. Fields of the order of 0.1 tesla (T) are feasible for a 40°C temperature rise above room ambient temperature. With supercooled coils, fields of 10 T have been obtained.

Figure 1.7 shows a transformer in its simplest form: two air coils that share a common flux. This flux diverges from the ends of the coil in all directions. It is not concentrated or confined. One of the coils is connected to the source and carries a current that establishes a magnetic field. The other coil is open-circuited. Notice that the flux lines are not all common to both coils. The difference between the two is the leakage flux; that is, *leakage flux* is the portion of the flux that does not link both coils.

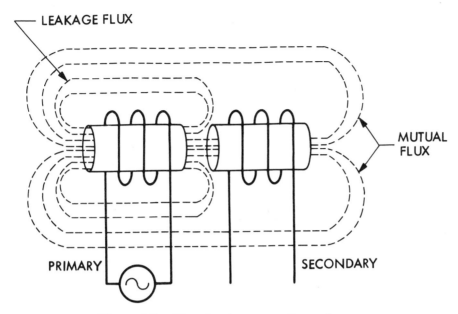

Figure 1.7. The simplest type of transformer.

1.1.2 Magnetic Core

Most materials are poor conductors of magnetic flux; they have low permeability. A vacuum has a permeability of 1.0, and nonmagnetic materials such as air, paper, and copper have permeabilities of the same order. There are a few materials such as iron, nickel, cobalt, and their alloys that have high permeabilities, sometimes ranging into the hundreds of thousands. These materials are used for practical cores.

To achieve an improvement over the air coil, a magnetic core can be introduced, as shown in Figure 1.8. In addition to its high permeability, the advantages of the magnetic core over the air core are that the magnetic path length (MPL) is well defined and the flux is essentially confined to the magnetic core except in the immediate vicinity of the winding.

1.1.3 Magnetic Core Magnetization Curve

The effect of exciting a completely demagnetized ferromagnetic material with an external magnetizing force H and increasing it slowly from zero is shown in Figure 1.9, where the resulting flux density is plotted as a function of the magnetizing force H. Note that at first the flux density increases very slowly up to point A, then increases very rapidly up to point B, and then almost stops increasing. Point B is called the *knee* of the curve. At point C the magnetic core material has saturated. From this point on, the slope of the curve is $B/H = 1$ gauss/oersted [G/Oe], the coil behaving as if it were air-cored. When the magnetic core is in hard saturation, the coil has the same permeability as air, or unity. Following the magnetization curve in Figure 1.9, Figures 1.10–1.12 show how the flux in the core is generated from the inside of the core to the outside until the core saturates.

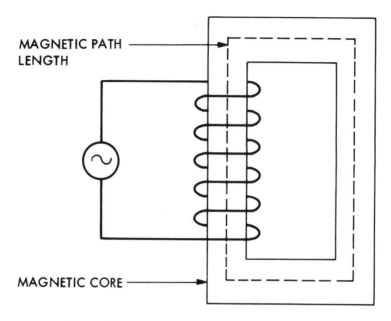

Figure 1.8. Introduction of a magnetic core.

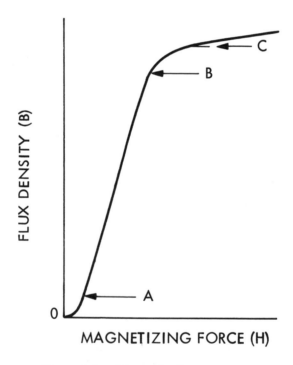

Figure 1.9. Magnetization curve.

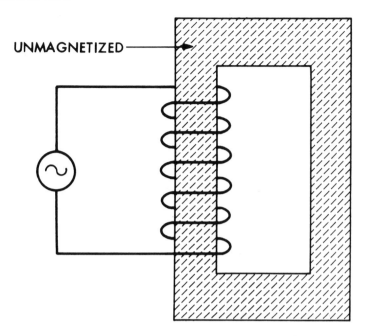

Figure 1.10. Demagnetized core.

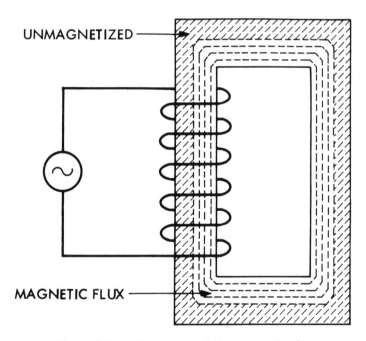

Figure 1.11. Core is partially magnetized.

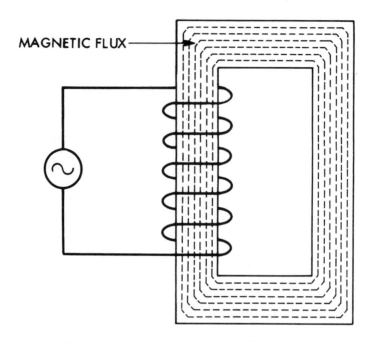

MAGNETIC FLUX

Figure 1.12. Magnetic core is saturated.

1.1.4 Hysteresis Loop

When the magnetic material is taken through a complete cycle of magnetization and demagnetization, the results are as shown in Figure 1.13. Starting with a neutral magnetic material, traversing the B-H loop starts at X. As H is increased, the flux density B increases along the dashed line to the saturation point, B_s. When H is now decreased and B plotted, the B-H loop transverses a path to B_r, where H is zero and the core is still magnetized. The flux here is called *remanent flux* and has a flux density B_r.

The magnetizing force H is now reversed in polarity to give a negative value. The magnetizing force required to reduce the flux B_r to zero is called the *coercive force* (H_c). When the core is forced into saturation, the *retentivity* (B_{rs}) is the remaining flux after saturation, and *coercivity* (H_{cs}) is the force required to reset to zero. Along the initial magnetization curve (X, the dashed line in Figure 1.13), B increases from the origin nonlinearly with H until the material saturates. In practice, the magnetization of a core in an excited transformer never follows this curve, because the core is never in the totally demagnetized state when the magnetizing force is first applied.

The hysteresis loop represents energy lost in the core. The best way to display the hysteresis loop is to use a dc current, because the intensity of the magnetizing force must be so slowly changed that no eddy currents are generated in the material. Only under this condition is the area inside the closed B-H loop indicative of the hysteresis loss. The enclosed area is a measure of energy lost in the core material during that cycle. In ac applications, this process is repeated continuously and the total hysteresis loss is dependent upon the frequency.

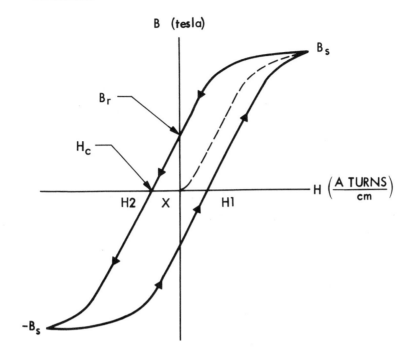

Figure 1.13. Hysteresis loop.

1.1.5 The Dynamic Hysteresis Loop

The hysteresis loss, however, is only part of the energy loss encountered in core material subjected to an alternating field. The changing flux also induces within the core material itself small electric currents known as *eddy currents*. The magnitude of these currents depends upon the frequency and flux density imposed by the application and upon the specific resistivity and thickness of the core material. Increasing the frequency or cycle rate between $+H$ and $-H$, we observe that the *B-H* loop widens as shown in Figure 1.14. The widening of the *B-H* loop is caused by eddy currents, primarily which are generated by the magnetic flux as it penetrates the laminations. The magnetic flux induces a voltage and causes a current to flow around the flux line as shown in Figure 1.15. The thicker the lamination, the higher the current.

Since the eddy currents are proportional to the voltage induced in the core material, the eddy current loss will be decreased when the voltage is decreased. The maximum voltage that can be induced in the core material is equal to the volts per turn of the core winding. The amount of current is a function of the induced voltage and the resistivity of the magnetic material. Eddy currents can be reduced by using thinner laminations and or a magnetic material with higher resistivity. With laminations that are half as thick, the flux in each lamination and the maximum voltage that generates eddy currents in each lamination are reduced to one-half their former values. The change of resistance in the eddy current path that takes place due to the thinner lamination is small compared with the change in voltage. The power changes with the square of the voltage and is linear with resistance.

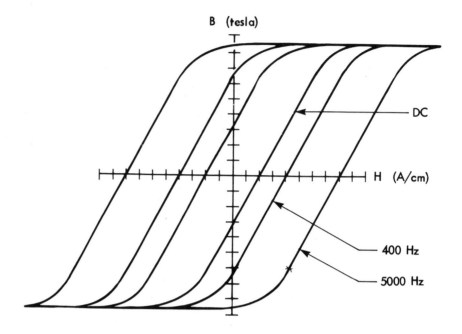

Figure 1.14. *B-H* loop operating at various frequencies.

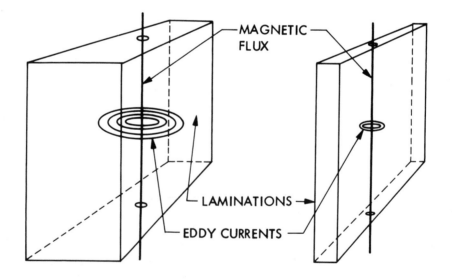

Figure 1.15. Eddy currents in both thick and thin laminations.

The total core loss is the sum of the hysteresis loss and eddy current loss. It is usually expressed in watts per kilogram of a material of a definite thickness for a core operating at a given flux density and frequency.

1.1.6 Permeability

In magnetics, *permeability* is the ability of a material to conduct flux. The magnitude of the permeability at a given induction is a measure of the ease with which a core material can be magnetized to that induction. It is defined as the ratio of the flux density B to the magnetizing force H. Manufacturers specify permeability in units of gauss per oersted (G/Oe).

$$\text{Permeability} = \mu = \frac{B}{H} \left[\frac{\text{gauss}}{\text{oersted}} \right] \tag{1.1}$$

Absolute permeability μ_0 in cgs units is unity (1 gauss/oersted) in a vacuum.

$$\text{cgs:} \quad \mu_0 = 1 \left[\frac{\text{gauss}}{\text{oersted}} \right] = \frac{\text{tesla}}{\text{oersted}} \times 10^4 \quad \leftarrow 10^{-4}\,? \tag{1.2}$$

$$\text{mks:} \quad \mu_0 = 0.4\pi \times 10^{-8} \left[\frac{\text{henry}}{\text{meter}} \right] \quad \text{or} \quad \left[\frac{\text{Wb}}{\text{A-m}} \right] \tag{1.3}$$

When B is plotted against H as in Figure 1.16, the resulting curve is called the *magnetization curve*. These curves are idealized. The magnetic material is totally demagnetized and is then subjected to

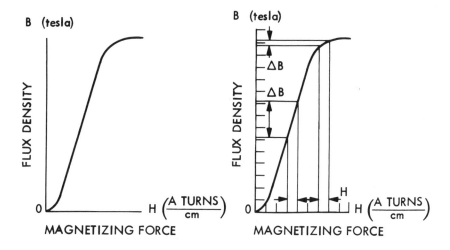

Figure 1.16. Magnetization curve.

a gradually increasing magnetizing force while the flux density is plotted. The slope of this curve at any given point gives the permeability at that point. Permeability can be plotted against a typical *B-H* curve as shown in Figure 1.17a. Permeability is not constant; therefore its value can be stated only at a given value of *B* or *H*.

There are many different kinds of permeability, and each is designated by a different subscript on the symbol μ.

μ_0 Absolute permeability, defined as the permeability in a vacuum.

μ_i Initial permeability, Figure 1.17b, the slope of the initial magnetization curve at the origin. It is measured at very small inductions.

μ_Δ Incremental permeability, Figure 1.17c, the slope of the magnetization curve for finite values of peak-to-peak flux density with superimposed dc magnetization.

μ_e Effective permeability. If a magnetic circuit is not homogeneous (i.e., contains an air gap), the effective permeability is the permeability of a hypothetical homogeneous (ungapped) structure of the same shape, dimensions, and reluctance that would give the inductance equivalent to the gapped structure.

μ_r Relative permeability, the permeability of a material relative to that of free space.

μ_n Normal permeability, Figure 1.17d, the ratio of B/H at any point of the curve.

μ_{max} Maximum permeability, Figure 1.17e, the slope of a straight line drawn from the origin tangent to the curve at its knee.

μ_p Pulse permeability, the ratio of peak B to peak H for unipolar excitation.

μ_m Material permeability, the slope of the magnetization curve measured at less than 50 G.

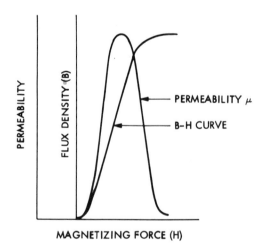

Figure 1.17a.   Variation of μ along the magnetization curve.

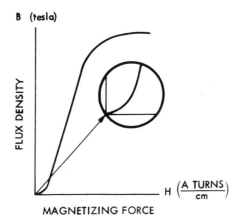

Figure 1.17b. Initial permeability.

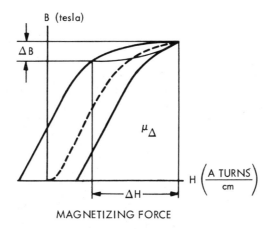

Figure 1.17c. Incremental permeability.

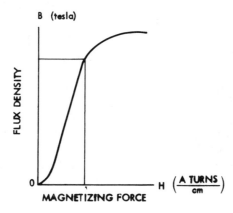

Figure 1.17d. Normal permeability.

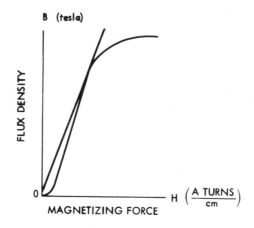

Figure 1.17e. Maximum permeability.

1.1.7 Reluctance

The flux produced in a given material by mmf depends on the material's "resistance" to flux, which is called its *reluctance R*. The reluctance of a core depends on the composition of the material and its physical dimensions and is similar in concept to electrical resistance. The relation between mmf, flux, and magnetic reluctance is analogous to the relation between emf, current, and resistance as shown in Figure 1.18.

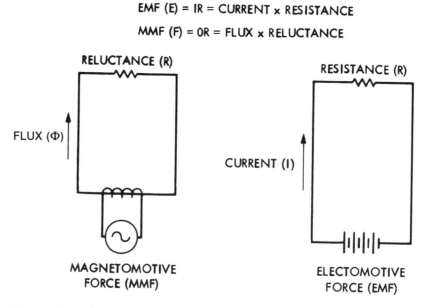

Figure 1.18. Illustrating the dual relation between magnetomotive force and electromotive force.

A poor conductor of flux has a high magnetic resistance R. The greater the reluctance, the higher the magnetomotive force required to obtain a given magnetic field.

The electrical resistance of a conductor is related to its length l, cross-sectional area A_w, and specific resistance p, which is the resistance per unit length. To find the resistance of a copper wire of any size or length, we merely multiply the resistivity by the length, and divide by the cross-sectional area:

$$R = \frac{pl}{A_w} \tag{1.4}$$

In the case of a magnetic circuit, $1/\mu$ is analogous to p and is called *reluctivity*. The *reluctance R_m* of a magnetic circuit is given by

$$R_m = \frac{MPL}{\mu_r \mu_0 A_c} \tag{1.5}$$

where MPL is the magnetic path length, cm
    A_c is the cross-sectional area of the core, cm^2
    μ_r is the permeability of the magnetic material *(or relative permeability?*
    μ_0 is the permeability of air

Figure 1.19 shows the magnetic path length MPL and the cross-sectional area A_c for an iron core.

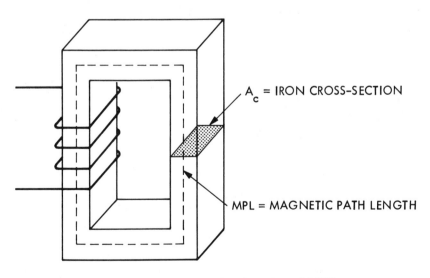

Figure 1.19. Magnetic core showing A_c and MPL.

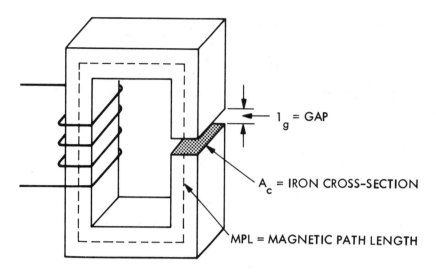

Figure 1.20.   A magnetic core with an air gap.

Thus, a high-permeability material is one that has a low reluctance for a given magnetic path length MPL and iron cross section A_c. If an air gap is included in a magnetic circuit as shown in Figure 1.20, which is otherwise composed of low-reluctivity material like iron, almost all the reluctance in the circuit will be that of the air gap, because the reluctivity of air is much greater than that of a magnetic material. Controlling the size of the air gap therefore controls the reluctance, for all practical purposes.

This can best be shown with an example. The total reluctance of the core is the sum of the iron reluctance and the air gap reluctance, in the same way that two series resistors are added in an electrical circuit. The equation for calculating the gap reluctance R_g is basically the same as the equation for calculating the reluctance of the magnetic material R_m. The difference is that the permeability of air is 1 and the gap length l_g is used in place of the magnetic path length MPL. This gives the equation

$$R_g = \left(\frac{1}{\mu_0}\right)\left(\frac{l_g}{A_c}\right) \tag{1.6}$$

But since $\mu_0 = 1$, this simplifies to

$$R_g = \frac{l_g}{\mu_0 A_c} \tag{1.7}$$

where l_g is the gap length, cm
$\quad$ A_c is the cross section of the core, cm^2
$\quad$ μ_0 is the permeability of air

· The total reluctance, R_t, for the core shown in Figure 1.20 is therefore

$$R_t = R_m + R_g \tag{1.8}$$

$$R_t = \frac{\text{MPL}}{\mu_r\mu_0 A_c} + \frac{l_g}{\mu_0 A_c} \tag{1.9}$$

where μ_r is the relative permeability, which is used exclusively with magnetic materials,

$$\mu_r = \frac{\mu}{\mu_0} = \frac{B}{\mu_0 H} \tag{1.10}$$

The magnetic material permeability μ_m is given by

$$\mu_m = \mu_r\mu_0 \tag{1.11}$$

The reluctance of the gap is higher than that of the iron even when the gap is very small. This is because even poor grades of silicon-iron have over 4000μ while the best nickel-iron has around 100,000μ so the total reluctance of the circuit depends more on the gap than on the iron.

After the total reluctance R_t has been calculated, the effective permeability μ_e can be calculated.

$$R_t = \frac{l_t}{\mu_e A_c} \tag{1.12}$$

$$l_t = l_g + \text{MPL} \tag{1.13}$$

where l_t is the total path length
μ_e is the effective permeability

$$R_t = \frac{l_t}{\mu_e A_c} = \frac{l_g}{\mu_0 A_c} + \frac{\text{MPL}}{\mu_0\mu_r A_c} \tag{1.14}$$

Simplifying yields

$$\frac{l_t}{\mu_e} = \frac{l_g}{\mu_0} + \frac{\text{MPL}}{\mu_0\mu_r} \tag{1.15}$$

Then

$$\mu_e = \frac{l_t}{l_g/\mu_0 + \text{MPL}/\mu_0\mu_r} \tag{1.16}$$

$$\mu_e = \frac{l_g + \text{MPL}}{l_g/\mu_0 + \text{MPL}/\mu_0\mu_r} \tag{1.17}$$

If $l_g \ll \text{MPL}$, multiply both sides of the equation by $\mu_r\mu_0\text{MPL}/\mu_r\mu_0\text{MPL}$. Then

$$\mu_e = \frac{\mu_0\mu_r}{1 + \mu_r(l_g/\text{MPL})} \tag{1.18}$$

The classic equation is

$$\mu_e = \frac{\mu_m}{1 + \mu_m(l_g/\text{MPL})} \tag{1.19}$$

As the gap is increased, so is the reluctance. For a given magnetomotive force, the flux density is controlled by the gap reluctance.

1.1.8 Magnetomotive Force (MMF) and Magnetizing Force (*H*)

There are two force functions commonly encountered in magnetics: magnetomotive force mmf and magnetizing force *H*. Magnetomotive force should not be confused with magnetizing force; the two are related as cause and effect. *Magnetomotive force* is given by the equation

$$\text{mmf} = 0.4\pi NI \quad \text{[gilberts]} \tag{1.20}$$

where N is the number of turns
    I is the current, amperes

whereas mmf is the force, *H* is a force field, or force per unit length:

$$H = \frac{\text{mmf}}{\text{MPL}} \quad \left[\frac{\text{gilberts}}{\text{cm}} = \text{oersteds}\right] \tag{1.21}$$

Substituting,

$$H = \frac{0.4\pi NI}{\text{MPL}} \quad [\text{oersteds}] \qquad (1.22)$$

where MPL = magnetic path length, cm.

If the flux ϕ is divided by the core area A_c, we get flux density B in lines per unit area:

$$B = \frac{\phi}{A_c} \qquad (1.23)$$

The flux density B in a magnetic medium due to the existence of a magnetizing force field H depends on the permeability of the medium and the intensity of the magnetic field.

$$B = \mu H \quad [\text{gauss}] \qquad maxwells/_{cm^2} \qquad (1.24)$$

The peak magnetizing current I_m for a wound core can be calculated from the following equation:

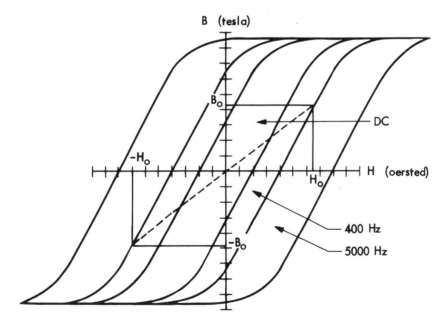

Figure 1.21. Typical *B-H* loops result from operating at different frequencies.

$$I_m = \frac{H_0 \, \text{MPL}}{0.4\pi N} \qquad\qquad (1.25)$$

where H_0 is the field intensity at the peak operating point.

To determine the magnetizing force H_0, use the manufacturer's core loss curves at the appropriate frequency and operating flux density B_0 as shown in Figure 1.21.

1.2 SELECTION OF MAGNETIC MATERIALS

Transformers used in static inverters, converters, and transformer-rectifier (T-R) supplies intended for aerospace and electronics industry power applications are usually of square loop tape toroidal design. The design of reliable, efficient, and lightweight devices for this use has been seriously hampered by the lack of engineering data describing the behavior of both the commonly used and more exotic core materials with higher-frequency square wave excitation.

A program has been carried out at JPL to develop these data from measurements of the dynamic *B-H* loop characteristics of the tape core materials presently available from various industry sources. Cores were produced in both toroidal and C forms and were tested in both ungapped (uncut) and gapped (cut) configurations. This section describes the results of this investigation.

1.2.1 Typical Operation

Transformers used for inverters, converters, and transformer-rectifier supplies operate from a power bus, which could be dc or ac. In some power applications, a commonly used circuit is a driven transistor switch arrangement such as that shown in Figure 1.22.

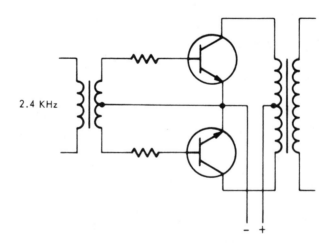

Figure 1.22. Typical driven transistor inverter.

One important consideration affecting the design of suitable transformers is that care must be taken to ensure that operation involves balance drive to the transformer primary. In the absence of balanced drive, a net dc current will flow in the transformer primary, which causes the core to saturate easily during alternate half-cycles. A saturated core cannot support the applied voltage, and, because of lowered transformer impedance, the current flowing in a switching transistor is limited mainly by its beta. The resulting high current, in conjunction with the transformer leakage inductance, results in a high-voltage spike during the switching sequence that could be destructive to the transistors. To provide balanced drive, it is necessary to exactly match the transistors for $V_{CE(SAT)}$ and beta, and this is not always sufficiently effective. Also, exact matching of the transistors is a major problem in the practical sense.

1.2.2 Material Characteristics

Many available core materials approximate the ideal square loop characteristic illustrated by the *B-H* curve shown in Figure 1.23. Representative dc *B-H* loops for commonly available core materials are shown in Figure 1.24. Other characteristics are tabulated in Table 1.1.

Many articles have been written about inverter and converter transformer design. Usually, their authors' recommendations represent a compromise among material characteristics such as those tabulated in Table 1.1 and displayed in Figure 1.24. These data are typical of commercially available core materials that are suitable for the particular application.

As can be seen, the material that provides the highest flux density (silicon) would result in the smallest component size, and this would influence the choice if size were the most important consideration. The type 78 material (see the 78% curve in Figure 1.24) has the lowest flux density. This results in the largest transformer, but, on the other hand, this material has the lowest coercive force and the lowest core loss of any core material available.

Usually, inverter transformer design is aimed at the smallest size with the highest efficiency and adequate performance under the widest range of environmental conditions. Unfortunately,

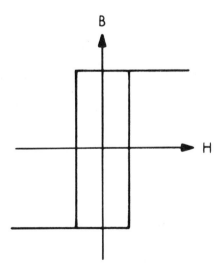

Figure 1.23.   Ideal square *B-H* loop.

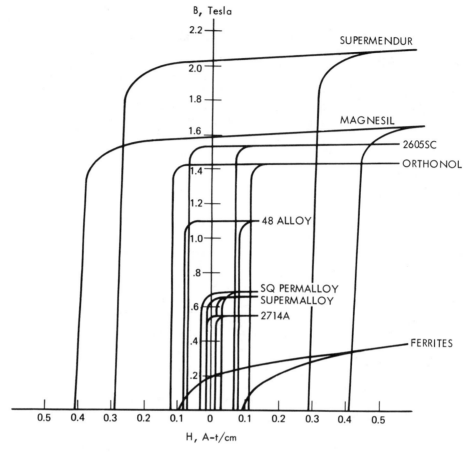

Figure 1.24. The typical dc *B-H* loops of magnetic materials. (*Courtesy of Magnetics Inc.*)

the core material that can produce the smallest size has the lowest efficiency, and the highest efficiency materials result in the largest size. Thus the trnasformer designer must make tradeoffs between allowable transformer size and the minimum efficiency that can be tolerated. The choice of core material will then be based upon achieving the best characteristic on the most critical or important design parameter and acceptable compromises on the other parameters.

After analysis of a number of designs, most engineers choose size rather than efficiency as the most important criterion and select an intermediate loss factor core material for their transformers. Consequently, as the frequency is increased, ferrites have become the most popular material.

1.2.3 Core Saturation Defined

To standardize the definition of saturation, several unique points on the *B-H* loop are defined as shown in Figure 1.25.

The straight line through $(H_0, 0)$ and (H_s, B_s) may be written as

$$B = \left(\frac{dB}{dH}\right)(H - H_0) \tag{1.26}$$

Table 1.1.   Magnetic core material characteristics

TRADE NAMES	COMPOSITION	* SATURATED FLUX DENSITY, tesla	DC COERCIVE FORCE, AMP–TURN / cm	SQUARENESS RATIO	** MATERIAL DENSITY, g / cm^3	CURIE TEMP. ^{0}C	WEIGHT FACTOR
Supermendur Permendur	49% Co 49% Fe 2% V	1.9–2.2	0.18–0.44	0.90–1.0	8.15	930	1.066
Magnesil Silectron Microsil Supersil	3% Si 97% Fe	1.5–1.8	0.5–0.75	0.85–0.75	7.63	750	1.00
Deltamax Orthonol 49 Sq Mu	50% Ni 50% Fe	1.4–1.6	0.125–0.25	0.94–1.0	8.24	500	1.079
Allegheny 4750 48 Alloy Carpenter 49	48% Ni 52% Fe	1.15–1.4	0.062–0.187	0.80–0.92	8.19	480	1.073
4–79 Permalloy Sp Permalloy 80 Sq Mu 79	79% Ni 17% Fe	0.66–0.82	0.025–0.82	0.80–1.0	8.73	460	1.144
Supermalloy	78% Ni 17% Fe 5% Mo	0.65–0.82	0.0037–0.01	0.40–0.70	8.76	400	1.148
Metglas: 2605SC	81% Fe 3.5% Si 13.5% B 2% C	1.5–1.6	.03–.08	.9–.98	7.32	370	.96
2714A	66% Co 4% Fe 15% Si 14% B 1% Ni	.5–.65	.008–.015	.9–.98	7.59	205	.995
Ferrites F N27 3C8	Mn Zn	0.45–0.50	0.25	0.30–0.5	4.8	250	0.629

*tesla = 10^4 Gauss
** g / cm^3 = 0.036 lb / in^3

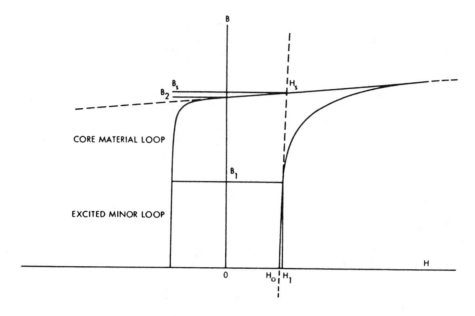

Figure 1.25. Defining the *B-H* loop.

The line through $(0, B_s)$ and (H_s, B_s) has essentially zero slope and may be written as

$$B = B_2 \approx B_s \qquad (1.27)$$

Equations (1.26) and (1.27) together define *saturation conditions* as follows:

$$B_s = \left(\frac{dB}{dH}\right)(H_s - H_0) \qquad (1.28)$$

Solving Equation (1.28) for H_s yields

$$H_s = H_0 + \frac{B_s}{\mu_0} \qquad (1.29)$$

where, by definition,

$$\mu_0 = \frac{dB}{dH}$$

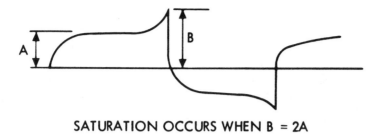

SATURATION OCCURS WHEN B = 2A

Figure 1.26.   Excitation current.

Saturation occurs by definition when the peak exciting current is twice the average exciting current as shown in Figure 1.26. Analytically this means that

$$H_{pk} = 2H_s \tag{1.30}$$

Solving Equation (1.26) for H_1, we obtain

$$H_1 = H_0 + \frac{B_1}{\mu_0} \tag{1.31}$$

To obtain the presaturation dc margin (ΔH), Equation (1.29) is subtracted from Equation (1.31):

$$\Delta H = H_s - H_1 = \frac{B_s - B_1}{\mu_0} \tag{1.32}$$

The actual unbalanced dc current must be limited to

$$I_{dc} \leqslant \frac{\Delta H \, l_m}{N} \quad \text{[amperes]} \tag{1.33}$$

where N is the number of turns
      l_m is the mean magnetic length

Combining Equations (1.32) and (1.33) gives

$$I_{dc} \leqslant \frac{(B_s - B_1)l_m}{\mu_0 N} \tag{1.34}$$

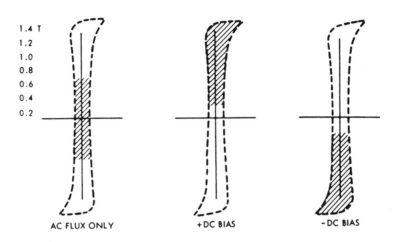

Figure 1.27.   *B-H* loop with dc bias.

As mentioned earlier, in an effort to prevent core saturation, the switching transistors are matched for beta and V_{CE}(SAT) characteristics. The effect of core saturation using an uncut or un-gapped core is shown in Figure 1.27, which illustrates the effect on the *B-H* loop traversed with a dc bias. Figure 1.28 shows typical *B-H* loops of 50-50 nickel-iron excited from an ac source with progressively reduced excitation; the vertical scale is 0.4 T/cm. It can be noted that the minor loop remains at one extreme position within the *B-H* major loop after reduction of excitation. The unfortunate effect of this random minor loop positioning is that when conduction again begins in the transformer winding after shutdown, the flux swing could begin from the extreme rather than from the normal zero axis. The effect of this is to drive the core into saturation, with the production of spikes that can destroy transistors.

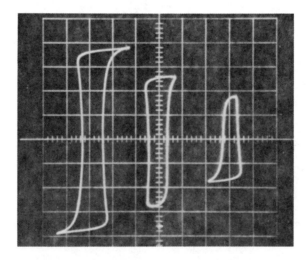

Figure 1.28.   Typical square loop material with ac excitation.

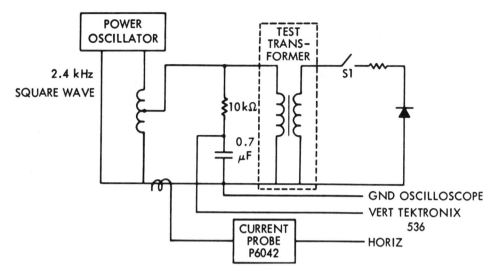

Figure 1.29. Dynamic *B-H* loop test fixture.

1.2.4 The Test Setup

The test fixture schematically illustrated in Figure 1.29 was built to effect comparison of dynamic *B-H* loop characteristics of various core materials. Cores were fabricated from various core materials in the basic core configuration designated No. 52029 for toroidal cores manufactured by Magnetics Inc. The materials used were those most likely to be of interest to designers of inverter or converter transformers. Test conditions are listed in Table 1.2.

Winding data were derived from the following:

$$N = \frac{V \times 10^4}{4.0 B_m f A_c} \qquad (1.35)$$

Table 1.2. Materials and test conditions

Core type*	Material	B_m, T	N_T	Frequency, kHz	l_m, cm
52029 (2A)	Orthonol	1.45	54	2.4	9.47
52029 (2D)	Sq. Permalloy	0.75	54	2.4	9.47
52029 (2F)	Supermalloy	0.75	54	2.4	9.47
52029 (2H)	48 Alloy	1.15	54	2.4	9.47
52029 (2H)	Magnesil	1.6	54	2.4	9.47

*Magnetics Inc. toroidal cores.

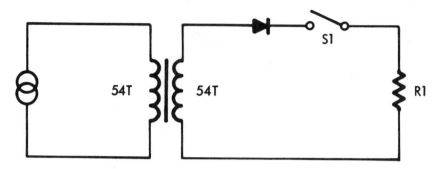

Figure 1.30. Implementing dc unbalance.

The test transformer represented in Figure 1.30 consists of 54-turn primary and secondary windings, with square wave excitation on the primary. Normally switch S1 is open. With switch S1 closed, the secondary current is rectified by the diode to produce a dc bias in the secondary winding.

Cores were fabricated from each of the materials by winding a ribbon of the same thickness on a mandrel of a given diameter. Ribbon termination was effected by welding in the conventional manner. The cores were vacuum impregnated, baked, and finished as usual.

Figures 1.31–1.35 show the dynamic *B-H* loops obtained for various core materials.

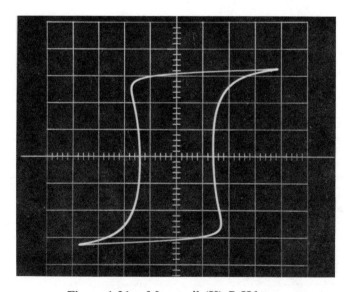

Figure 1.31. Magnesil (K) *B-H* loop.

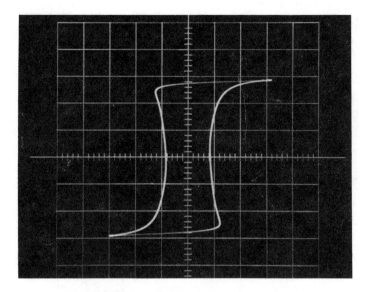

Figure 1.32.   Orthonol (A) *B-H* loop.

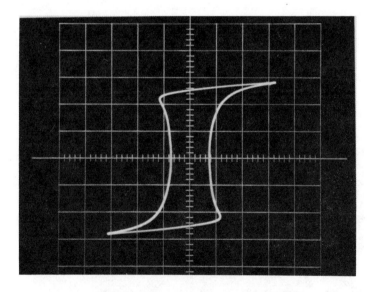

Figure 1.33.   48 Alloy (H) *B-H* loop.

Figure 1.36 shows a composite of all the *B-H* loops. In each of these, switch S1 was in the open position, so there was no dc bias applied to the core and windings.

Figures 1.37–1.41 show the dynamic *B-H* loop patterns obtained for various core materials when the test conditions included a sequence in which switch S1 was open, then closed, and then opened. It is apparent from these data that with a small amount of dc bias the minor dynamic *B-H* loop can traverse the major *B-H* loop from saturation to saturation. Note that after the dc bias has

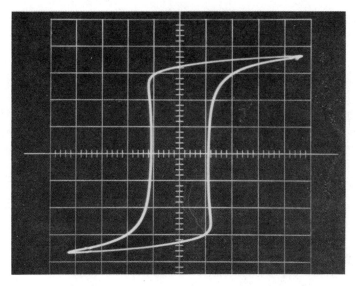

Figure 1.34. Sq. Permalloy (P) *B-H* loop.

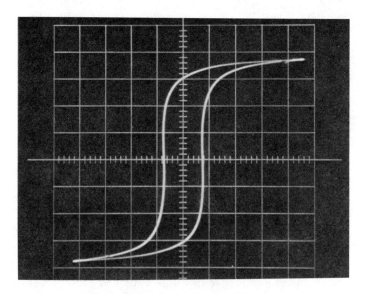

Figure 1.35. Supermalloy (F) *B-H* loop.

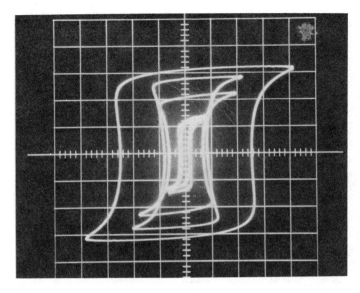

Figure 1.36. Composite 52029 (2K), (A), (H), (P), and (F) *B-H* loops.

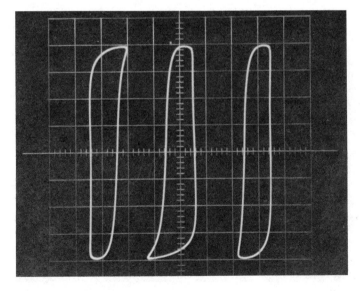

Figure 1.37. Magnesil (K) *B-H* loop with and without dc bias.

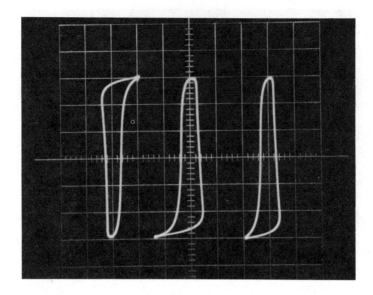

Figure 1.38. Orthonol (A) *B-H* loop with and without dc bias.

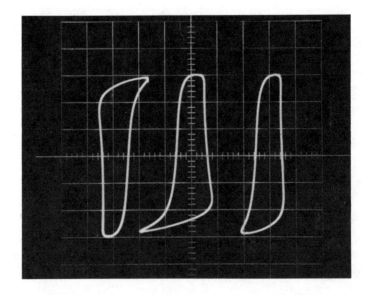

Figure 1.39. 48 Alloy (H) *B-H* loop with and without dc bias.

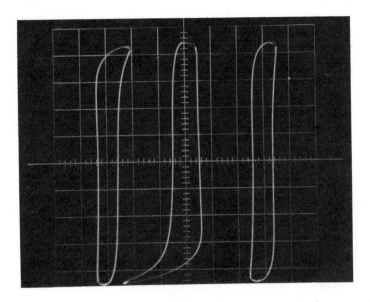

Figure 1.40. Sq. Permalloy (P) *B-H* loop with and without dc bias.

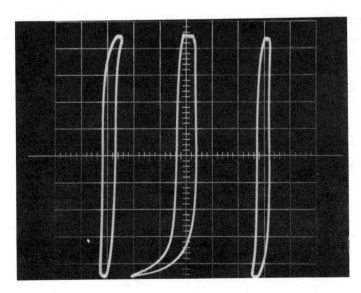

Figure 1.41. Supermalloy (F) *B-H* loop with and without dc bias.

removed the minor *B-H* loops remained shifted to one side or the other. Because of the ac coupling of the integrator to the oscilloscope, the photographs in these figures do not present a complete picture of what really happens during the flux swing.

1.2.5 Core Saturation Theory

The domain theory of the nature of magnesium is based on the assumption that all magnetic materials consist of individual molecular magnets. These minute magnets are capable of movement within the material. When a magnetic material is in its unmagnetized state, the individual magnetic particles are arranged at random and effectively neutralize each other. An example of this is shown in Figure 1.42, where the tiny magnetic particles are arranged in a disorganized manner. (The north poles are represented by the darkened ends of the magnetic particles.) When a material is magnetized, the individual particles are aligned or oriented in a definite direction (Figure 1.43).

The degree of magnetization of a material depends on the degree of alignment of the particles. The external magnetizing force can continue to affect the material up to the point of saturation, the point at which essentially all of the domains are lined up in the same direction.

In a typical toroidal core, the effective air gap is less than 10^{-6} cm. Such a gap is negligible in comparison to the ratio of mean length to permeability. If the toroid were subjected to a strong magnetic field (enough to saturate), essentially all of the domains would line up in the same direction. If suddenly the field were removed at B_m, the domains would remain lined up and be magnetized along that axis. The amount of flux density that remains is called the *residual flux, B_r.* The result of this effect was shown earlier in Figures 1.37–1.41.

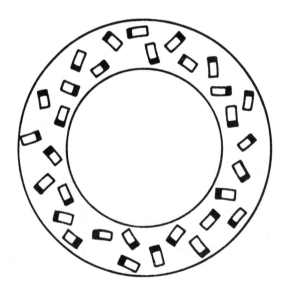

Figure 1.42. Unmagnetized material.

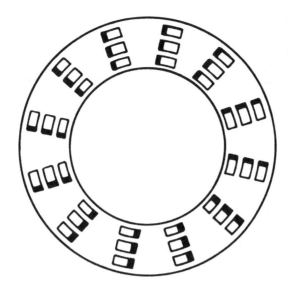

Figure 1.43.   Magnetized material.

1.2.6 Air Gap

An air gap introduced into the core has a powerful demagnetizing effect, resulting in a "shearing over" of the hysteresis loop and a considerable decrease in permeability of high-permeability materials. Direct current excitation follows the same pattern. However, the core bias is considerably less affected than the magnetization characteristics by the introduction of a small air gap. The magnitude of the air gap effect also depends on the length of the mean magnetic path and on the characteristics of the uncut core. For the same air gap, the decrease in permeability will be less with a greater magnetic flux path but more pronounced in a high-permeability core with a low coercive force.

1.2.7 Effect of Gapping

Figure 1.44 shows a comparison of a typical toroidal core *B-H* loop without and with a gap. The gap increases the effective length of the magnetic path. When voltage E is impressed across primary winding N_1 of a transformer, the resulting current i_m will be small because of the highly inductive circuit shown in Figure 1.45. For a particular core size, maximum inductance occurs when the air gap is minimum.

When S1 is closed, an unbalanced dc current flows in the N_2 turns, and the core is subjected to a dc magnetizing force, resulting in a flux density that may be expressed as

$$B_{dc} = \frac{0.4\pi N I_{dc} \times 10^{-4}}{l_g + l_m/\mu_r} \quad \text{[teslas]} \tag{1.36}$$

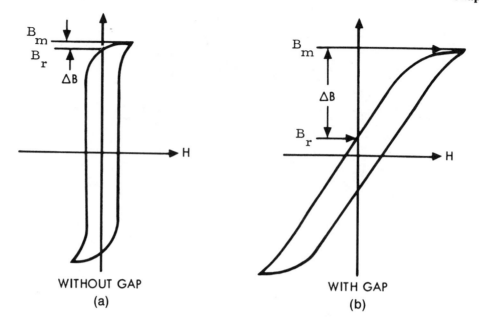

Figure 1.44. Air gap increases the effective length of the magnetic path.

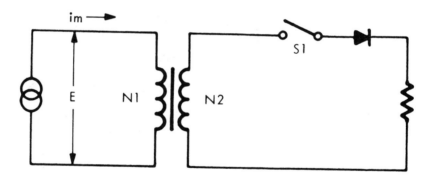

Figure 1.45. Implementing dc unbalance.

In converter and inverter design, this is augmented by the ac flux swing, which is

$$B_{ac} = \frac{E \times 10^4}{KfA_cN} \quad [\text{teslas}] \tag{1.37}$$

If the sum of B_{dc} and B_{ac} shifts operation above the maximum operating flux density of the core material, the incremental permeability (μ_{ac}) is reduced. This lowers the impedance and in-

Figure 1.46.   Typical cut toroid.

creases the flow of magnetizing current i_m. This can be remedied by introducing an air gap into the core assembly, which causes a decrease in dc magnetization in the core. However, the size of the air gap that can be incorporated has a practical limitation, since the air gap lowers impedance, resulting in increased magnetizing current i_m, which is inductive. The resultant voltage spikes produced by such currents apply a high stress to the switching transistors and may cause failure. This stress can be minimized by tight control of lapping and etching of the gap to keep the gap to a minimum.

From Figure 1.44, it can be seen that the B-H curves depict maximum flux density B_m and residual flux B_r for ungapped and gapped cores, and that the useful flux swing is designated ΔB, which is the difference between B_m and B_r. It will be noted in Figure 1.44(a) that B_r approaches B_m but in Figure 1.44(b) there is a much greater ΔB between them. In either case, when excitation voltage is removed at the peak of the excursion of the B-H loop, flux falls to the B_r point. It is apparent that introducing an air gap reduces B_r to a lower level and increases the useful flux density. Thus insertion of an air gap in the core eliminates, or markedly reduces, the voltage spikes produced by the leakage inductance due to the transformer saturation.

Two types of core configurations were investigated in the ungapped and gapped states. Figure 1.46 shows the type of toroidal core that was cut, and Figure 1.47 shows the type of C core that was cut. As conventionally fabricated, toroidal cores are virtually gapless. To increase the gap, the cores were physically cut in half, and the cut edges were lapped, acid etched to remove cut debris, and banded to form the cores. A minimum air gap on the order of less than 25 μm was established.

As will be noted from Figures 1.48–1.52, which show the B-H loops of the uncut and cut cores, the results obtained indicated that the effect of gapping was the same for both the C cores and the toroidal cores subjected to testing. It will be noted, however, that gapping of the toroidal cores produced a lowered squareness characteristic for the B-H loop as shown in Table 1.3; these data were obtained from Figures 1.48–1.52. ΔH values extracted from the same figures, as shown in Figure 1.53, are tabulated in Table 1.4.

A direct comparison of cut and uncut cores was made electrically by means of two different test circuits. The magnetic material used in this branch of the test was Orthonol. The operating fre-

Figure 1.47. Typical cut C core.

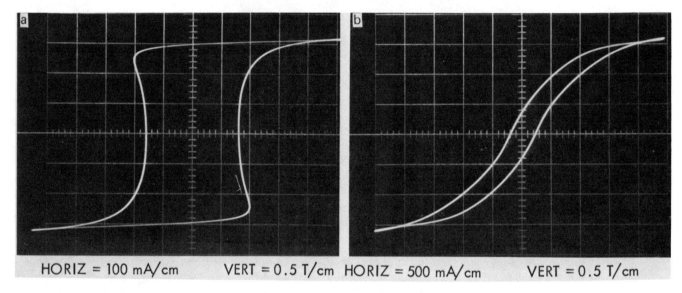

Figure 1.48. Magnesil 52029 (2K) *B-H* loops. (*a*) Uncut; (*b*) cut.

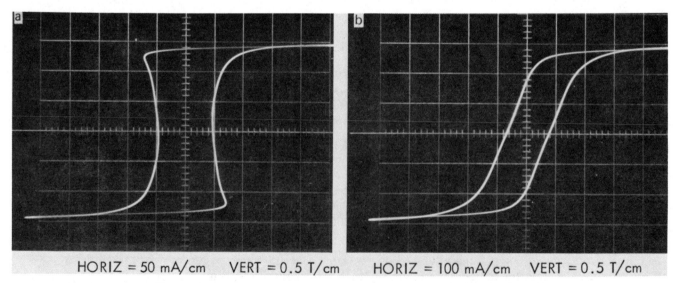

Figure 1.49. Orthonol 52029 (2A) *B-H* loop. (*a*) Uncut; (*b*) cut.

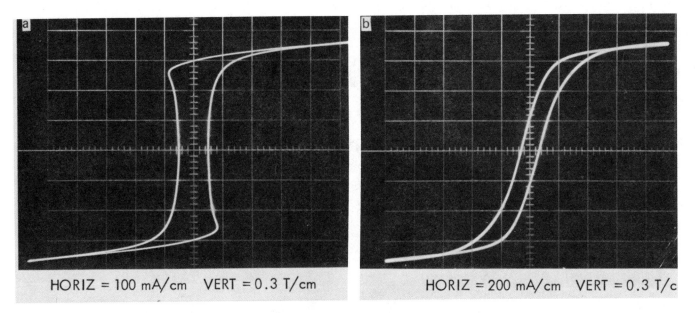

Figure 1.50. 48 Alloy 52029 (2H) *B-H* loop. (*a*) Uncut; (*b*) cut.

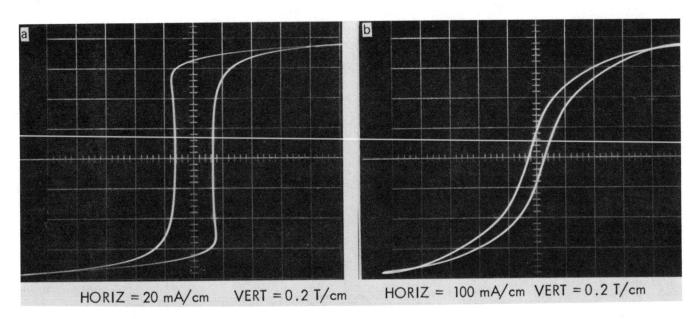

Figure 1.51. Sq. Permalloy 52029 (2D) *B-H* loop. (*a*) Uncut; (*b*) cut.

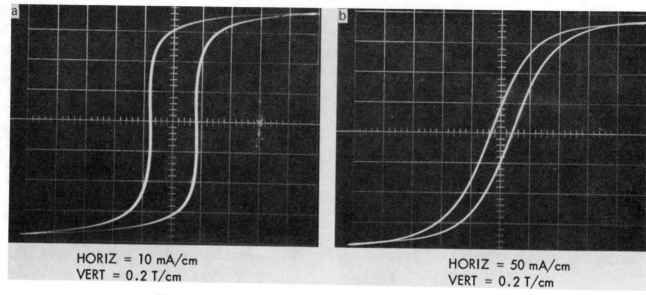

HORIZ = 10 mA/cm
VERT = 0.2 T/cm

HORIZ = 50 mA/cm
VERT = 0.2 T/cm

Figure 1.52. Supermalloy 52029 (2F) *B-H* loop. (*a*) Uncut; (*b*) cut.

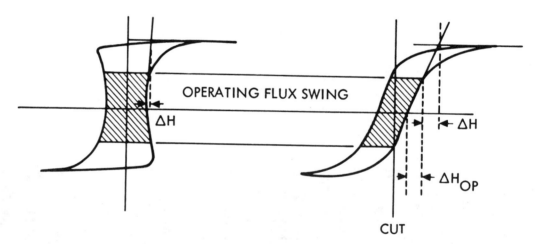

OPERATING FLUX SWING

ΔH

ΔH

ΔH$_{OP}$

CUT

Figure 1.53. Defining ΔH and ΔH_{OP}.

Table 1.3. Comparing B_r/B_m on uncut and cut cores

Code	Material	Uncut B_r/B_m	Cut B_r/B_m
(A)	Orthonol	0.96	0.62
(D)	Molypermalloy	0.86	0.21
(K)	Magnesil	0.93	0.22
(F)	Supermalloy	0.81	0.24
(H)	48 Alloy	0.83	0.30

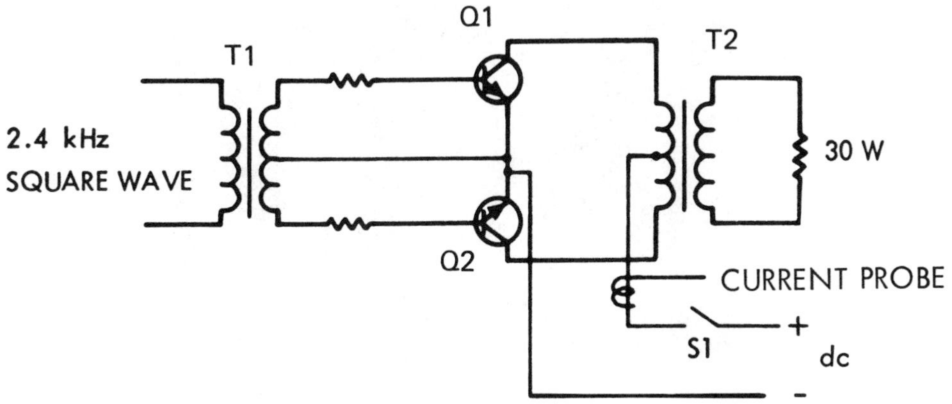

Figure 1.54. Inverter inrush current measurement.

Table 1.4. Comparing ΔH and ΔH_{OP} on uncut and cut cores

				Amp-turns/cm			
				Uncut		Cut	
Material	B_m, T	B_{ac}, T	B_{dc}, T	ΔH_{OP}	ΔH	ΔH_{OP}	ΔH
Orthonol	1.44	1.15	0.288	0.0125	0.0	0.895	0.178
48 Alloy	1.12	0.89	0.224	0.0250	0.0	1.60	0.350
Sq. Permalloy	0.73	0.58	0.146	0.01	0.005	0.983	0.178
Supermalloy	0.68	0.58	0.136	0.0175	0.005	0.491	0.224
Magnesil	1.54	1.23	0.31	0.075	0.025	7.15	1.78

quency was 2.4 kHz, and the flux density was 0.6 T. The first test circuit, shown in Figure 1.54, was a driven inverter operating into a 30-W load, with the transistors operating into and out of saturation. Drive was applied continuously. S1 controls the supply voltage to Q1 and Q2.

With switch S1 closed, transistor Q1 was turned on and allowed to saturate. This applied E-V_C(SAT) across the transformer winding. Switch S1 was then opened. The flux in transformer T2 then dropped to the residual flux density B_r. Switch S1 was closed again. This was done several times in succession to catch the flux in an additive direction. Figures 1.55 and 1.56 show the inrush current measured at the center tap of T2.

It will be noted in Figure 1.55 that the uncut core saturated and that inrush current was limited only by circuit resistance and transistor beta. Figure 1.56 shows that saturation did not occur in the case of the cut core. The high inrush current and transistor stress were thus virtually eliminated.

The second test circuit arrangement is shown in Figure 1.57. The purpose of this test was to excite a transformer and measure the inrush current using a current probe. A square wave power oscillator was used to excite transformer T2. Switch S1 was opened and closed several times to catch the flux in an additive direction. Figures 1.58 and 1.59 show inrush current for a cut and uncut core, respectively.

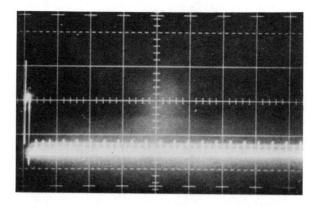

Figure 1.55.   Typical inrush current of an uncut core in a driven inverter.

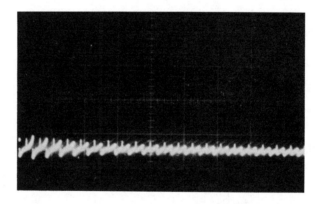

Figure 1.56.   Typical inrush current of a cut core in a driven inverter.

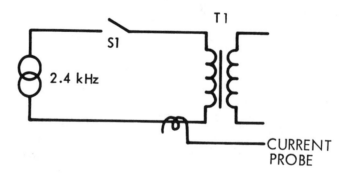

Figure 1.57.   T-R supply current measurement.

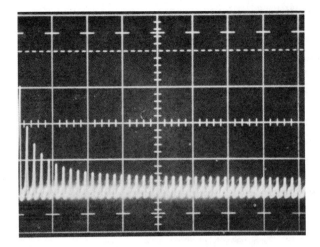

Figure 1.58.   Typical inrush current of an uncut core operating from an ac source.

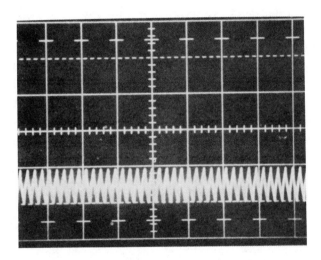

Figure 1.59.   Typical inrush current of a cut core in a transformer-rectifier.

A small amount of air gap, less than 25 μm, has a powerful effect on the demagnetizing force but little effect on core loss. This small air gap decreases the reisdual magnetism by "shearing over" the hysteresis loop, which eliminates the problem of the core tending to remain saturated.

A typical example of the merits of the cut core occurred in the checkout of a Mariner space-craft. During the checkout of a prototype science package, a large (8 A, 200 μs) turn-on transient was observed. The normal running current was 0.06 A, fused with a parallel-redundant 1/8-A fuse as required by the Mariner Mars 1971 design philosophy. With the 8-A inrush current, the 1/8-A fuses were easily blown. This did not happen on every turn-on but only when the core would "latch up" in the wrong direction for turn-on. Upon inspection, the transformer turned out to be a 50-50 nickel-iron toroid. The design was changed from a toroidal core to a cut core with a 25-μm air gap. The new design was completely successful in eliminating the 8-A turn-on transient.

1.2.8 A New Core Configuration

A new configuration has been developed for transformers that combines the protective feature of a gapped core with the much lower magnetizing current requirement of an uncut core. The uncut core functions under normal operating conditions, and the cut core takes over during abnormal conditions to prevent high switching transients and their potentially destructive effect on the transistors.

This configuration is a composite of cut and uncut cores assembled together concentrically, with the uncut core nested within the cut core. The uncut core has high permeability and thus requires a very small magnetizing current. On the other hand, the cut core has a low permeability and thus requires a much higher magnetization current. The uncut core is designed to operate at a flux density that is sufficient for normal operation of the converter. The uncut core may saturate under the abnormal conditions previously described. The cut core than takes over and supports the applied voltage so that excessive current does not flow. In a sense, it acts like a ballast resistor in some circuits to limit current flow to a safe level.

Figures 1.60 and 1.61 show the magnetization curves for an uncut core and a composite core of the same material at the same flux density. The much lower B_r characteristic of the composite compared to the uncut core is readily apparent.

The desired features of the composite core can be obtained more economically by using different materials for the cut and uncut portions of the core. It was found that when the design required high nickel (4/79), the cut portion could be low nickel (50/50), and because low nickel has twice as high a flux density as high nickel the core was made 66% high nickel and 33% low nickel.

Figure 1.62 shows cut and uncut cores that have been impregnated to bond the ribbon layers together. The uncut core was first trimmed to fit within the inner diameter of the cut core by peeling off a wrap or two of the ribbon steel. The two cores are assembled into a composite core (Fig. 1.63, left), which is then placed in an aluminum box and sealed (Fig. 1.63, right).

To ensure uniform characteristics for gapped cores, a gap dimension of 50 μm is recommended, because variations produced by thermal cycling will not affect this gap greatly. In the composite core, the gap is obtained by inserting a sheet of paper or film material between the core ends during banding. The same protective feature can be accomplished in transformers with laminationed cores. When laminations are stacked by interleaving them one by one, the result will be a

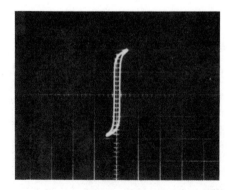

Figure 1.60.   The uncut core excited at 0.2 T/cm.

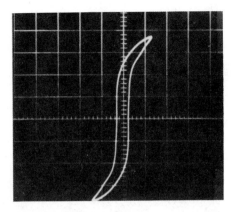

Figure 1.61.   Both cut and uncut cores excited at 0.2 T/cm.

Figure 1.62.   Uncut and cut cores before assembly to form a composite core. The uncut core has had some of the outer wrapping removed so it will fit within the inner diameter of the cut core.

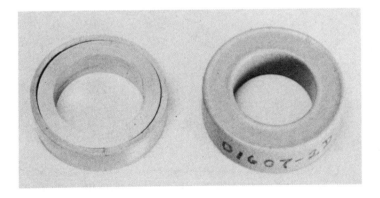

Figure 1.63.   Cores after assembly (left) and in final form, sealed in an aluminum box (right).

minimum air gap as shown in Figure 1.64 by the squareness of the *B-H* loop. Shearing over of the *B-H* loop or decreasing the residual flux as shown in Figure 1.65 is accomplished by butt joining half the laminations in the core cross section, which introduces a small additional air gap.

Table 1.5 is a compilation of composite cores manufactured by Magnetics, Inc., alongside their standard dimensional equivalent cores. Also included in Table 1.5 is the cores' area product A_p, and the core geometry K_g which is discussed in Chapter 2.

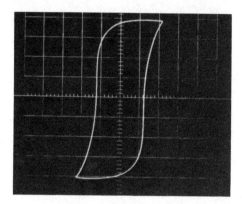

Figure 1.64. *B-H* loop for core with interleaved laminations.

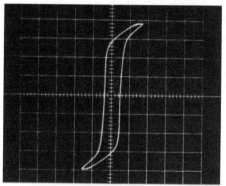

Figure 1.65. *B-H* loop for laminated core with half the laminations butt-stacked.

1.2.9 Summary

Low-loss tape-wound toroidal core materials that have a very square hysteresis characteristic (*B-H* loop) have been used extensively in the design of spacecraft transformers. Due to the squareness of the *B-H* loops of these materials, transformers designed with them tend to saturate quite easily. As a result, large voltage and current spikes, which cause undue stress on the electronic circuitry, can occur. Saturation occurs when there is any unbalance in the ac drive to the transformer or when any dc excitation exists. Also, due to the square characteristic, a high residual flux state (high B_r)

Table 1.5.   Area products of composite cores and equivalent standard cores

Composite Core	Standard Core	WaAc (cm^4)	Kg (cm^5)
01605-2D	52000	0.0728	0.00105
01754-2D	52002	0.144	0.00171
01755-2D	52076	0.285	0.00661
01609-2D	52061	0.389	0.00744
01756-2D	52106	0.439	0.00948
01606-2D	52094	0.603	0.0221
01761-2D	52318	0.779	0.0260
01757-2D	52029	1.090	0.0256
01760-2D	52188	1.152	0.0512
02153-2D	52181	1.220	0.0407
01758-2D	52032	1.455	0.0431
01607-2D	52026	2.180	0.0874
01966-2D	52030	2.337	0.0635
01759-2D	52038	2.910	0.140
01608-2D	52035	4.676	0.206
01623-2D	52425	5.255	0.262
01624-2D	52169	7.13	0.418

A_c = 66% Square Permalloy 4/79.

A_c = 33% Orthonol 50/50.

I_g = 2 mil Kapton.

Source: Magnetics, Inc.

may remain when excitation is removed. Reapplication of excitation in the same direction may cause deep saturation, and an extremely large current spike, limited only by source impedance and transformer winding resistance, can result. This can produce catastrophic failure.

With the introduction of a small (less than 25-μm) air gap into the core, the problems described above can be avoided while retaining the low-loss properties of the materials. The air gap has the effect of "shearing over" the *B-H* loop of the material so that the residual flux state is low and the margin between operating flux density and saturation flux density is high. The air gap thus has a powerful demagnetizing effect upon the square loop materials. Properly designed transformers using cut toroid or C core square-loop materials will not saturate upon turn-on and can tolerate a certain amount of unbalanced drive or dc excitation.

It must be emphasized, however, that because of the nature of the material and the small size of the gap, extreme care and control must be taken in performing the gapping operation. Otherwise the desired shearing effect will not be achieved, and the low-loss properties will be lost. The cores must be very carefully cut, lapped, and etched to provide smooth, residue-free surfaces. Reassembly must be performed with equal care.

BIBLIOGRAPHY

Brown, A. A. et al. (1969). *Cyclic and Constant Temperature Aging Effects on Magnetic Materials for Inverters and Converters,* NASA CR-(L-80001). National Aeronautics and Space Administration, Washington, D.C., June 1969.

Design Manual Featuring Tape Wound Cores, TWC-300, Magnetics Inc., Butler, Pennsylvania, 1962.

Flight Projects, Space Programs Summary 37-64, Vol. 1, Jet Propulsion Laboratory, Pasadena, California, July 31, 1970, p. 17.

Frost, R. M. et al. (1969). *Evaluation of Magnetic Materials for Static Inverters and Converters,* NASA CR-1226, National Aeronautics and Space Administration, Washington, D.C., February 1969.

Lee, R. (1958). *Electronic Transformers and Circuits, 2nd ed., John Wiley & Sons, New York.*

Nordenberg, H. M. (1964). Electronic Transformers. Reinhold, New York.

Platt, S. (1958). *Magnetic Amplifiers: Theory and Application,* Prentice-Hall, Englewood Cliffs, New Jersey.

Technical Data on Arnold Tape-Wound Cores, TC-101A. Arnold Engineering, Marengo, Illinois, 1960.

Chapter 2

Transformer Design Tradeoffs

2.1 INTRODUCTION

The conversion process in power electronics requires the use of transformers, components that are frequently the heaviest and bulkiest item in the conversion circuits. Transformers also have a significant effect upon the overall performance and efficiency of the system. Accordingly, their design has an important influence on overall system weight, power conversion efficiency, and cost. Because of the interdependence and interaction of parameters, judicious tradeoffs are necessary to achieve design optimization.

2.2 POWER-HANDLING ABILITY

Manufacturers have for years assigned numeric codes to their cores to indicate their power-handling ability. This method assigns to each core a number called the *area product A_p* that is the product of its window area W_a and core cross section area A_c. These numbers are used by core suppliers to summarize dimensional and electrical properties in their catalogs. They are available for laminations, C cores, pot cores, powder cores, ferrite toroids, and toroidal tape-wound cores.

The author has developed additional relationships between the A_p numbers and current density J for a given regulation and temperature rise. The area product A_p is a length dimension to the fourth power (l^4), whereas volume is a length dimension to the third power (l^3), and surface area A_t is a length dimension to the second power (l^2). Straight-line relationships have been developed for A_p and volume, A_p and surface area A_t, and A_p and weight. These are discussed at appropriate points throughout this volume.

The regulation (see Chapter 3) and power-handling ability of a core are related to its inherent core geometry. The core geometry coefficient K_g is a relatively new concept, and magnetic core manufacturers are just beginning to use it.

Because of their significance, the area product A_p and core geometlry K_g are treated extensively in this handbook. A great deal of other information is also presented for the convenience of the designer. Much of the material is in tabular form to assist the designer in making the tradeoffs best suited for the particular application in a minimum amount of time.

2.3   RELATIONSHIP OF K_g TO POWER TRANSFORMER REGULATION CAPABILITY

Although most transformers are designed for a given temperature rise, they can also be designed for a given regulation. The regulation and power-handling ability of a core are related to the two constants K_g and K_e by the following equation:

$$\alpha = \frac{P_t}{2K_g K_e} \quad [\%]$$

$$\alpha = \text{regulation} \quad [\%]$$

The constant K_g is determined by the core geometry which may be related by the following equations:

$$K_g = \frac{W_a A_c^2 K_u}{\text{MLT}} \quad [\text{cm}^5]$$

The constant K_e is determined by the magnetic and electrical operating conditions, which may be related by the following equation:

$$K_e = 0.145 K_f^2 f^2 B_m^2 \times 10^{-4}$$

where

$$K_f, \text{waveform coefficient} = \begin{cases} 4.0 & \text{for a square wave} \\ 4.44 & \text{for a sine wave} \end{cases}$$

2.4 RELATIONSHIP OF A_p TO TRANSFORMER POWER-HANDLING CAPABILITY

The power-handling capability of a core is related to its area product by the equation

$$A_p = \frac{P_t \times 10^4}{K_f B_m f K_u J} \quad [\text{cm}^4]$$

where K_f is the waveform coefficient

$$K_f = \begin{cases} 4.0 & \text{for a square wave} \\ 4.44 & \text{for a sine wave} \end{cases}$$

From the above it can be seen that factors such as flux density, frequency of operation, window utilization factor K_u, which defines the maximum space which may be occupied by the copper in the window, and the current density J, which is related to temperature rise, all have an influence on the transformer area product.

2.5 THE AREA PRODUCT A_p AND ITS RELATIONSHIPS

The area product A_p of a core is the product of the available window area W_a of the core in square centimeters (cm^2) multiplied by the effective cross-sectional area A_c in square centimeters (cm^2), which may be stated as

$$A_p = W_a A_c \quad [\text{cm}^4]$$

Figures 2.1–2.5 show in outline form five transformer core types that are typical of those shown in suppliers' catalogs.

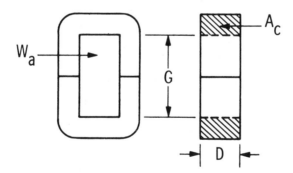

Figure 2.1. C core.

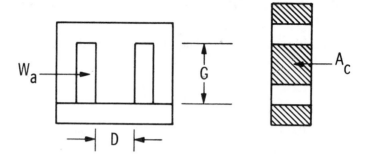

Figure 2.2. EI lamination.

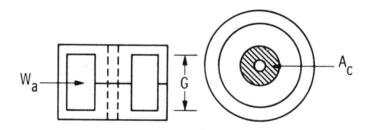

Figure 2.3. Pot core.

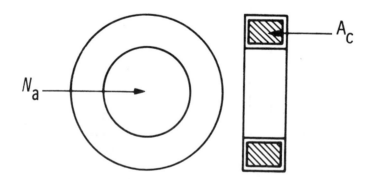

Figure 2.4. Tape-wound toroidal core.

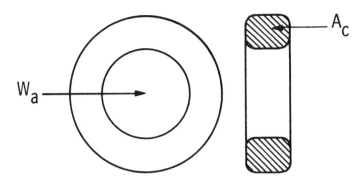

Figure 2.5.   Powder core.

Table 2.1.   Core configuration constants

Core	Losses	K_j (25°C)	K_j (50°C)	(x)	K_s	K_w	K_v
Pot core	$P_{cu} = P_{fe}$	433	632	−0.17	33.8	48.0	14.5
Powder core	$P_{cu} \gg P_{fe}$	403	590	−0.12	32.5	58.8	13.1
Lamination	$P_{cu} = P_{fe}$	366	534	−0.12	41.3	68.2	19.7
C core	$P_{cu} = P_{fe}$	323	468	−0.14	39.2	66.6	17.9
Single-coil	$P_{cu} \gg P_{fe}$	395	569	−0.14	44.5	76.6	25.6
Tape-wound core	$P_{cu} = P_{fe}$	250	365	−0.13	50.9	82.3	25.0

$$J = K_j A_p^{(x)} \qquad\qquad A_t = K_s A_p^{0.50}$$
$$W_t = K_w A_p^{0.75} \qquad\qquad \text{Vol} = K_v A_p^{0.75}$$

There is a unique relationship between the area product A_p characteristic number for transformer cores and several other important parameters that must be considered in transformer design.

Table 2.1 was developed using the least-squares curve fit from the data obtained in Tables 2.2–2.7. The relationships of the area product A_p with volume, surface area, current density, and weight for pot cores, powder cores, laminations, C cores, and tape-wound cores are presented in detail in the table footnotes.

Table 2.2. Powder core characteristics

		1	2	3	4	5		6	7	8
						N				$I = \sqrt{\dfrac{W}{\Omega}}$
		Core	A_t, cm^2	A_p, cm^4	MLT, cm		AWG	Ω(50°C)	P_Σ	
1		55051	6.569	0.0432	2.16	86		0.215	0.216	1.00
							25			
2		55121	11.24	0.139	2.74	160		0.513	0.369	0.848
							25			
3		55848	15.69	0.264	2.97	257		0.897	0.519	0.761
							25			
4		55059	20.02	0.460	3.45	316		1.27	0.657	0.719
							25			
5		55894	28.32	0.997	4.61	351		1.87	0.924	0.703
							25			
6		55586	44.24	1.83	4.32	902		4.69	1.46	0.558
							25			
7		55071	40.68	1.95	4.80	656		3.70	1.34	0.602
							25			
8		55076	46.91	2.44	4.88	815		4.71	1.55	0.574
							25			
9		55083	61.05	4.53	6.07	959		6.84	2.00	0.541
							25			
10		55090	81.58	8.06	6.66	1372		10.8	2.68	0.498
							25			
11		55439	79.37	8.33	7.62	959		8.49	2.60	0.553
							25			
12		55716	91.32	9.32	6.50	1684		13.0	3.00	0.480
							25			
13		55110	112.4	13.65	7.00	2125		17.8	3.72	0.457
							25			

Copper loss ≫ iron loss

Definitions for Table 2.2

Information given is listed by column as:

1. Manufacturer part number (Magnetics Inc.). Arnold Engineering also manufactures powder cores.
2. Surface area calculated from Figure 2.22
3. Area product-effective iron area × window area
4. Mean length turn
5. Total number of turns and wire size using a window utilization factor $K_u = 0.40$
6. Resistance of the wire at 50°C
7. Watts loss is based on Figure 7.2 for a ΔT of 25°C with a room ambient of 25°C surface dissipation times the transformer surface area; total loss is P_{cu}
8. Current calculated from columns 6 and 7

9	10	11	12	13	14		15	16
$\Delta T(25°C)$ $J = I/cm^2$	$\Omega(75°C)$	P_Σ	$I = \sqrt{\dfrac{W}{\Omega}}$	$\Delta T(50°C)$ $J = I/cm^2$	Weight		Volume, cm^3	A_c, cm^2
					fe	cu		
617	0.236	0.503	1.46	899	3.1	2.71	1.39	0.113
522	0.563	0.861	1.23	762	6.8	6.3	3.11	0.196
469	0.985	1.211	1.11	683	10	11.3	5.07	0.232
443	1.39	1.533	1.05	647	16	16.3	7.28	0.327
433	2.06	2.16	1.02	631	36	23.2	12.4	0.639
344	5.15	3.40	0.812	500	35	59.9	23.3	0.458
371	4.07	3.13	0.877	540	47	47.4	21.0	0.666
353	5.17	3.61	0.814	518	52	61.0	25.7	0.670
333	7.50	4.68	0.790	487	92	86.0	39.1	1.06
307	11.8	6.26	0.728	449	131	140	59.5	1.32
341	9.32	6.08	0.807	497	182	109	58.1	1.95
296	14.3	7.00	0.699	431	133	170	69.0	1.24
282	19.6	8.68	0.665	410	176	226	93.4	1.44

9. Current density calculated from columns 5 and 8
10. Resistance of the wire at 75°C
11. Watts loss is based on Figure 7.2 for a ΔT of 50°C with a room ambient of 25°C surface dissipation times the transformer surface area; total loss is P_{cu}
12. Current calculated from columns 10 and 11
13. Current density calculated from columns 5 and 12
14. Effective core weight for molypermalloy powder plus copper, in grams
15. Transformer volume calculated from Figure 2.6
16. Core effective cross section

Table 2.3. Pot core characteristics

	1	2	3	4	5		6	7	8
					N				$I = \sqrt{\dfrac{W}{\Omega}}$
	Core	A_t, cm^2	A_p, cm^4	MLT, cm		AWG	$\Omega(50°C)$	P_Σ	
1	9 × 5	2.93	0.0065	1.85	25	30	0.175	0.098	0.529
2	11 × 7	4.35	0.0152	2.2	37	30	0.309	0.130	0.458
3	14 × 8	6.96	0.0393	2.8	74	30	0.787	0.208	0.363
4	18 × 11	11.3	0.114	3.56	143	30	1.934	0.339	0.296
5	22 × 13	17.0	0.246	4.4	207	30	3.46	0.510	0.271
6	26 × 16	23.9	0.498	5.2	96	25	0.592	0.717	0.778
7	30 × 19	32.8	1.016	6.0	144	25	1.024	0.984	0.693
8	36 × 22	44.8	2.01	7.3	189	25	1.636	1.34	0.639
9	47 × 28	76.0	5.62	9.3	345	25	3.81	2.28	0.547
10	59 × 36	122.0	13.4	12.0	608	25	8.65	3.66	0.459

Copper loss = iron loss

Definitions for Table 2.3

Information given is listed by column as:

1. Manufacturer part number (Siemens). Other manufacturers of similar pot cores include Magnetics Inc., Ferroxcube, Allen-Bradley, Indiana General, and TDK Ferrites.
2. Surface area calculated from Figure 2.22
3. Area product-effective iron area × window area
4. Mean length turn
5. Total number of turns and wire size using a window utilization factor $K_u = 0.40$
6. Resistance of the wire at 50°C
7. Watts loss is based on Figure 7.2 for a ΔT of 25°C with a room ambient of 25°C surface dissipation times the transformer surface area; total loss is equal to $2P_{cu}$
8. Current calculated from columns 6 and 7

9	10	11	12	13	14		15	16
$\Delta T(25°C)$ $J = I/cm^2$	$\Omega(75°C)$	P_Σ	$I = \sqrt{\dfrac{W}{\Omega}}$	$\Delta T(50°C)$ $J = I/cm^2$	Weight		Volume, cm^3	A_c, cm^2
					fe	cu		
1044	0.192	0.230	0.774	1527	0.8	0.32	0.367	0.10
904	0.339	0.304	0.670	1322	1.7	0.38	0.662	0.16
716	0.864	0.487	0.531	1048	3.2	0.98	1.35	0.25
584	2.12	0.791	0.432	853	6.0	2.37	2.78	0.43
535	3.80	1.190	0.396	782	13	4.30	5.17	0.63
479	0.650	1.67	1.13	696	21	7.5	8.65	0.94
427	1.12	2.30	1.01	622	36	12.9	13.9	1.36
394	1.79	3.14	0.937	577	57	20.8	22.0	2.01
337	4.18	5.32	0.798	492	125	48.0	48.6	3.12
283	9.50	8.54	0.670	413	270	109	98.3	4.85

9. Current density calculated from columns 5 and 8
10. Resistance of the wire at 75°C
11. Watts loss is based on Figure 7.2 for a ΔT of 50°C with a room ambient of 25°C surface dissipation times the transformer surface area; total loss is equal to $2P_{cu}$.
12. Current calculated from columns 10 and 11
13. Current density calculated from columns 5 and 12
14. Effective core weight for ferrite plus copper, in grams
15. Transformer volume calculated from Figure 2.6
16. Core effective cross section

Table 2.4. Lamination characteristics

	Core	A_t, cm²	A_p, cm⁴	MLT, cm	N	AWG	Ω(50°C)	P_Σ	$I = \sqrt{\dfrac{W}{\Omega}}$
		1	2	3	4	5	6	7	8
1	EE-3031	4.07	0.0088	1.72	90	30	0.58	0.123	0.323
2	EE-2829	6.53	0.0228	2.33	147	30	1.30	0.199	0.276
3	E1-187	14.2	0.108	3.20	314	30	3.82	0.432	0.237
4	EE-2425	23.3	0.293	5.08	498	30	9.61	0.714	0.192
5	EE-2627	38.5	0.906	5.79	245	25	1.68	1.22	0.602
6	E1-375	46.2	1.23	6.30	350	25	2.62	1.43	0.522
7	E1-50	53.2	1.75	7.09	263	25	2.21	1.73	0.625
8	E1-21	62.1	2.36	7.57	372	25	3.34	1.98	0.544
9	E1-625	83.2	4.29	8.84	503	25	5.27	2.70	0.505
10	E1-75	120.0	8.89	10.6	211	25	0.826	3.90	1.54
11	E1-87	163.0	16.5	12.3	296	20	1.34	5.28	1.40
12	E1-100	213.0	28.1	14.5	386	20	2.07	6.90	1.29
13	E1-112	270.0	44.9	16.0	492	20	2.91	8.76	1.23
14	E1-125	333.0	68.7	17.7	625	20	4.09	10.8	1.15
15	E1-138	403.0	107.0	19.5	740	20	5.33	13.0	1.10
16	E1-150	473.0	143.00	21.2	893	20	6.99	15.5	1.05
17	E1-175	742.0	263.0	24.7	1080	20	9.85	21.1	1.034
18	E1-36	649.0	324.0	26.5	1701	20	16.6	23.3	0.836
19	E1-19	1069.0	601.0	31.7	2886	20	33.8	32.8	0.696

Copper loss = iron loss

Definitions for Table 2.4

Information given is listed by column as:

1. Manufacturer part number (Magnetic Metals). Other manufacturers of laminated cores include Arnold Engineering, Thomas and Skinner, Tempel Steel Co., and Magnetics Inc.
2. Surface area calculated from Figure 2.23
3. Area product-effective iron area × window area
4. Mean length turn on one bobbin
5. Total number of turns and wire size for one bobbin using a window utilization factor $K_u = 0.40$
6. Resistance of the wire at 50°C
7. Watts loss is based on Figure 7.2 for a ΔT of 25°C with a room ambient of 25°C surface dissipation times the transformer surface area; total loss is equal to $2P_{cu}$
8. Current calculated from columns 6 and 7

9	10	11	12	13	14		15	16
$\Delta T(25°C)$ $J = I/cm^2$	$\Omega(75°C)$	P_Σ	$I = \sqrt{\dfrac{W}{\Omega}}$	$\Delta T(50°C)$ $J = I/cm^2$	Weight		Volume, cm^3	A_c, cm^2
					fe	cu		
638	0.645	0.288	0.472	932	1.02	1.02	0.651	0.0502
546	1.43	0.464	0.403	795	2.19	1.59	1.35	0.0907
469	4.19	1.01	0.347	685	7.09	3.08	4.34	0.204
380	10.5	1.67	0.281	555	15.5	9.06	9.22	0.363
371	1.85	2.84	0.876	540	45.8	15.5	19.1	0.816
322	2.87	3.34	0.762	470	49.7	24.7	25.3	0.816
385	2.43	4.04	0.912	562	90.6	31.7	36.8	1.45
335	3.66	4.62	0.793	489	99.3	41.0	39.2	1.45
312	5.79	6.30	0.737	455	179	44.4	60.0	2.27
296	0.906	9.10	2.24	432	312	105	104.0	3.27
270	1.48	12.3	2.04	393	481	135	164.0	4.45
249	2.27	16.1	1.88	363	712	241	246.0	5.81
237	3.19	20.4	1.79	344	1029	342	350.0	7.34
222	4.49	25.3	1.68	324	1414	460	481.0	9.07
213	5.85	30.2	1.61	310	1880	680	629.0	11.6
203	7.67	36.3	1.54	296	2457	906	829.0	13.1
199	10.8	49.3	1.51	291	3906	1273	1312.0	17.8
161	18.3	54.5	1.22	235	3575	2355	1654.0	15.3
134	37.1	76.5	1.015	196	4889	3805	2875.0	17.8

9. Current density calculated from columns 5 and 8
10. Resistance of the wire at 75°C
11. Watts loss is based on Figure 7.2 for a ΔT of 50°C with a room ambient of 25°C surface dissipation times the transformer surface area; total loss is equal to $2P_{cu}$
12. Current calculated from columns 10 and 11
13. Current density calculated from columns 5 and 12
14. Effective core weight for silicon plus copper, in grams
15. Transformer volume calculated from Figure 2.7
16. Core effective cross section (thickness, 0.014), square stack

Table 2.5. C-core characteristics

	Core	A_t, cm^2	A_p, cm^4	MLT, cm	N AWG	Ω(50°C)	P_Σ	$I = \sqrt{\dfrac{W}{\Omega}}$
		1	2	3	4	5	6	7	8
1	AL-2	20.9	0.265	3.55	662	8.93	0.627	0.187
						30			
2	AL-3	23.9	0.410	4.18	662	10.5	0.717	0.185
						30			
3	AL-5	33.6	0.767	4.59	946	16.5	1.01	0.174
						30			
4	AL-6	37.5	1.011	5.23	946	18.8	1.13	0.172
						30			
5	AL-124	45.3	1.44	5.50	1317	27.5	1.36	0.157
						30			
6	AL-8	63.4	2.31	5.74	221	0.482	1.90	1.404
						20			
7	AL-9	69.0	3.09	6.38	221	0.535	2.07	1.39
						20			
8	AL-10	74.5	3.85	7.01	221	0.588	2.24	1.38
						20			
9	AL-12	87.0	4.57	7.09	278	0.748	2.61	1.32
						20			
10	AL-135	93.7	5.14	7.36	325	0.908	2.81	1.24
						20			
11	AL-78	98.1	6.07	7.01	312	0.831	2.94	1.33
						20			
12	AL-18	118	7.92	7.61	510	1.47	3.55	1.10
						20			
13	AL-15	120	9.07	8.05	386	1.18	3.58	1.23
						20			
14	AL-16	127	10.8	8.89	386	1.30	3.80	1.20
						20			
15	AL-17	142	14.4	10.3	386	1.51	4.25	1.185
						20			
16	AL-19	159	18.0	10.8	511	2.10	4.77	1.065
						20			
17	AL-20	182	22.6	11.5	511	2.23	5.46	1.106
						20			
18	AL-22	202	28.0	11.5	637	2.78	6.05	1.043
						20			
19	AL-23	220	34.9	12.7	637	3.07	6.60	1.036
						20			
20	AL-24	245	40.0	12.0	948	4.32	7.35	0.922
						20			

Copper loss = iron loss

Definitions for Table 2.5

Information given is listed by column as:

1. Manufacturer part number (Arnold Engineering Co.). Other manufacturers of C cores include Magnetic Metals, National Magnetics, and Magnetics Inc.
2. Surface area calculated from Figure 2.24
3. Area product-effective iron area × window area
4. Mean length turn on one bobbin
5. Total number of turns and wire size for two bobbins using a window utilization factor $K_u = 0.40$
6. Resistance of the wire at 50°C
7. Watts loss is based on Figure 7.2 for a ΔT of 25°C with a room ambient of 25°C surface dissipation times the transformer surface area; total loss is equal to $2P_{cu}$

9	10	11	12	13	14		15	16
$\Delta T(25°C)$			$I = \sqrt{\dfrac{W}{\Omega}}$	$\Delta T(50°C)$	Weight		Volume,	
$J = I/cm^2$	$\Omega(75°C)$	P_Σ		$J = I/cm^2$	fe	cu	cm^3	A_c, cm^2
370	9.81	1.46	0.273	538	12.2	11.1	7.14	0.265
365	11.5	1.67	0.269	531	18.1	13.1	8.92	0.410
345	18.1	2.35	0.255	503	31.3	20.5	14.06	0.539
341	20.6	2.63	0.253	490	41.7	23.4	16.88	0.716
310	30.2	3.17	0.229	452	46.6	34.2	22.50	0.716
271	0.529	4.44	2.05	395	67.9	60.0	35.66	0.806
268	0.587	4.83	2.03	391	89.2	66.6	41.62	1.077
266	0.646	5.22	2.01	387	110.0	73.2	47.55	1.342
255	0.821	6.09	1.93	371	111.0	93.2	61.38	1.26
240	0.997	6.56	1.81	345	114.0	113.0	69.63	1.26
256	0.912	6.87	1.94	374	155.0	103.0	62.83	1.34
211	1.61	8.26	1.60	308	138.0	183.0	94.79	1.25
237	1.30	8.40	1.79	346	205.0	147.0	94.43	1.80
233	1.43	8.89	1.76	340	235.0	162.0	104.95	2.15
228	1.66	9.94	1.73	333	314.0	188.0	124.94	2.87
205	2.31	11.1	1.55	299	328.0	261.0	155.44	2.87
213	2.45	12.7	1.61	310	437.0	278.0	187.08	3.58
201	3.05	14.1	1.52	293	489.0	346.0	212.04	3.58
200	3.37	15.4	1.51	291	612.0	382.0	244.67	4.48
178	4.74	17.1	1.35	259	552.0	538.0	280.91	3.58

8. Current calculated from columns 6 and 7
9. Current density calculated from columns 5 and 8
10. Resistance of the wire at 75°C
11. Watts loss is based on Figure 7.2 for a ΔT of 50°C with a room ambient of 25°C surface dissipation times the transformer surface area; total loss is equal to $2P_{cu}$
12. Current calculated from columns 10 and 11
13. Current density calculated from columns 5 and 12
14. Effective core weight for silicon plus copper, in grams
15. Transformer volume calculated from Figure 2.8
16. Core effective cross section

Table 2.6. Single-coil C-core characteristics

	1	2	3	4	5		6	7	8
	Core	A_t, cm^2	A_p, cm^4	MLT, cm	N	AWG	$\Omega(50°C)$	P_Σ	$I = \sqrt{\dfrac{W}{\Omega}}$
1	AL-2	24.6	0.265	4.47	83		0.138	0.737	2.31
2	AL-3	27.6	0.410	5.10	83	20	0.158	0.828	2.28
3	AL-5	38.1	0.767	5.42	119	20	0.238	1.14	2.18
4	AL-6	41.9	1.011	6.06	119	20	0.266	1.26	2.17
5	AL-124	51.8	1.44	6.56	175	20	0.426	1.55	1.90
6	AL-8	72.8	2.31	7.06	255	20	0.669	2.18	1.80
7	AL-9	78.4	3.09	7.69	255	20	0.728	2.35	1.79
8	AL-10	83.9	3.85	8.33	255	20	0.788	2.52	1.78
9	AL-12	101.0	4.57	9.00	327	20	1.09	3.03	1.66
10	AL-135	110.0	5.14	9.50	370	20	1.31	3.30	1.58
11	AL-78	110.0	6.08	8.15	406	20	1.23	3.30	1.63
12	AL-18	142.0	7.87	7.51	564	20	2.14	4.26	1.41
13	AL-15	136.0	9.07	10.1	444	20	1.66	4.08	1.56
14	AL-16	143.0	10.8	10.7	444	20	1.77	4.29	1.55
15	AL-17	158.0	14.4	12.0	444	20	1.97	4.74	1.55
16	AL-19	182.0	18.1	13.0	563	20	2.71	5.46	1.41
17	AL-20	205.0	22.6	13.6	563	20	2.84	6.15	1.47
18	AL-22	228.0	28.0	13.6	704	20	3.56	6.84	1.38
19	AL-23	246.0	35.0	15.9	704	20	3.89	7.38	1.37
20	AL-24	282.0	40.0	14.6	1026	20	5.57	8.46	1.23

Definitions for Table 2.6

Information given is listed by column as:

1. Manufacturer part number (Arnold Engineering Co.). Other manufacturers of single-coil C cores include Magnetic Metals, National Magnetics, and Magnetics Inc.
2. Surface area calculated from Figure 2.25
3. Area product-effective iron area × window area
4. Mean length turn on one bobbin
5. Total number of turns and wire size for a single bobbin using a window utilization factor $K_u = 0.40$
6. Resistance of the wire at 50°C
7. Watts loss is based on Figure 7.2 for a ΔT of 25°C with a room ambient of 25°C surface dissipation times the transformer surface area; total loss is P_{cu}
8. Current calculated from columns 6 and 7

9	10	11	12	13	14		15	16
$\Delta T(25°C)$ $J = I/\text{cm}^2$	$\Omega(75°C)$	P_Σ	$I = \sqrt{\dfrac{W}{\Omega}}$	$\Delta T(50°C)$ $J = I/\text{cm}^2$	Weight		Volume, cm^3	A_c, cm^2
					fe	cu		
445	0.151	1.72	3.37	651	12.2	16.9	10.7	0.264
441	0.173	1.93	3/34	644	18.1	19.3	12.5	0.406
422	0.262	2.67	3.19	615	31.3	29.2	19.7	0.539
420	0.292	2.93	3.16	611	41.7	32.6	21.9	0.716
368	0.468	3.63	2.78	537	46.6	52.1	30.8	0.716
348	0.734	5.10	2.63	508	67.9	81.7	53.5	0.806
346	0.799	5.49	2.62	505	89.2	89.0	59.5	1.08
345	0.866	5.87	2.60	502	110.0	96.4	65.4	1.34
321	1.20	7.07	2.42	468	111.0	134.4	92.1	1.26
306	1.43	7.70	2.32	447	114.0	159.0	107.0	1.26
316	1.35	7.70	2.38	460	155.0	150.0	81.3	1.34
272	2.35	9.94	2.05	396	138.0	260.0	147.0	1.25
302	1.83	9.52	2.28	440	205.0	203.0	136.0	1.80
300	1.94	10.0	2.27	438	235.0	216.0	147.0	2.15
299	2.20	11.1	2.24	433	314.0	241.0	168.0	2.87
274	2.97	12.7	2.06	399	328.0	332.0	212.0	2.87
284	3.12	14.4	2.14	414	437.0	348.0	259.0	3.58
267	3.91	16.0	2.02	390	489.0	435.0	294.0	3.58
265	4.27	17.2	2.00	387	612.0	479.0	326.0	4.48
238	6.11	19.7	1.79	346	552.0	680.0	401.0	3.58

9. Current density calculated from columns 5 and 8
10. Resistance of the wire at 75°C
11. Watts loss is based on Figure 7.2 for a ΔT of 50°C with a room ambient of 25°C surface dissipation times the inductor surface area; total loss is P_{cu}
12. Current calculated from columns 10 and 11
13. Current density calculated from columns 5 and 12
14. Effective core weight plus copper, in grams
15. Transformer volume calculated from Figure 2.9
16. Core effective cross section

Table 2.7. Tape-wound toroidal core characteristics

	1	2	3	4	5		6	7	8
					N				$I = \sqrt{\dfrac{W}{\Omega}}$
	Core	A_t, cm^2	A_p, cm^4	MLT, cm		AWG	$\Omega(50°C)$	P_Σ	
1	52402	7.26	0.0100	2.05	302		2.35	0.218	0.215
						30			
2	52153	8.29	0.0196	2.22	302		2.54	0.249	0.221
						30			
3	52107	11.1	0.0201	2.21	606		5.09	0.333	0.180
						30			
4	52403	13.5	0.0267	2.30	621		5.43	0.405	0.193
						30			
5	52057	17.4	0.0659	2.53	1017		9.78	0.522	0.163
						30			
6	52000	15.2	0.0787	2.70	606		6.22	0.456	0.191
						30			
7	52063	20.7	0.132	2.85	1017		11.0	0.621	0.167
						30			
8	52002	21.8	0.144	2.88	1114		12.2	0.654	0.163
						30			
9	52007	27.6	0.380	3.87	982		14.4	0.828	0.169
						30			
10	52167	31.5	0.516	4.23	1000		16.1	0.945	0.171
						30			
11	52094	30.4	0.592	4.47	1017		17.3	0.912	0.162
						30			
12	52004	46.1	0.725	4.02	315		0.469	1.38	1.20
						20			
13	52032	56.5	1.46	4.65	315		0.543	1.69	1.25
						20			
14	52026	61.0	2.18	5.28	315		0.616	1.83	1.22
						20			
15	52038	65.9	2.91	5.97	315		0.697	1.98	1.19
						20			
16	52035	88.9	4.68	6.33	505		1.19	2.67	1.06
						20			

Copper loss = iron loss

Definitions for Table 2.7

Information given is listed by column as:

1. Manufacturer part number (Magnetics Inc.). Other manufacturers of tape-wound cores are Arnold Engineering and Magnetic Metals.
2. Surface area calculated from Figure 2.22
3. Area product-effective iron area × window area
4. Mean length turn
5. Total number of turns and wire size using a window utilization factor $K_u = 0.40$
6. Resistance of the wire at 50°C
7. Watts loss is based on Figure 7.2 for a ΔT of 25°C with a room ambient of 25°C surface dissipation times the transformer surface area; total loss is equal to $2P_{cu}$
8. Current calculated from columns 6 and 7

9	10	11	12	13	14		15	16
$\Delta T(25°C)$ $J = I/cm^2$	$\Omega(75°C)$	P_Σ	$I = \sqrt{\dfrac{W}{\Omega}}$	$\Delta T(50°C)$ $J = I/cm^2$	Weight		Volume, cm^3	A_c, cm^2
					fe	cu		
425	2.58	0.508	0.313	619	0.63	3.12	1.42	0.022
436	2.80	0.580	0.322	636	1.31	3.29	1.71	0.043
357	5.59	0.777	0.263	520	0.80	6.84	2.63	0.022
381	5.96	0.945	0.281	556	0.88	9.52	3.48	0.022
322	10.7	1.22	0.238	471	2.05	13.1	4.98	0.043
378	6.82	1.06	0.278	550	3.73	7.97	3.99	0.086
331	12.1	1.45	0.244	483	4.47	14.4	6.20	0.086
323	13.4	1.53	0.239	472	4.62	16.0	6.72	0.086
334	15.8	1.93	0.246	487	14.5	17.7	9.84	0.257
338	17.6	2.21	0.250	494	20.9	19.0	11.9	0.343
321	19.0	2.13	0.237	468	21.8	21.0	12.2	0.386
234	0.515	3.23	1.77	341	13.4	56.8	21.3	0.171
240	0.596	3.95	1.82	351	29.8	63.7	27.8	0.343
235	0.676	4.27	1.77	342	44.7	71.3	32.8	0.514
230	0.765	4.61	1.74	334	59.6	79.4	38.3	0.686
204	1.3	6.22	1.55	298	71.5	138.0	59.0	0.686

(continued)

9. Current density calculated from columns 5 and 8
10. Resistance of the wire at 75°C
11. Watts loss is based on Figure 7.2 for a ΔT of 50°C with a room ambient of 25°C surface dissipation times the transformer surface area; total loss is equal to $2P_{cu}$
12. Current calculated from columns 10 and 11
13. Current density calculated from columns 5 and 12
14. Effective core weight plus copper, in grams
15. Transformer volume calculated from Figure 2.6
16. Core effective cross section

Table 2.7 *(continued)*

		1	2	3	4	5		6	7	8
		Core	A_t, cm^2	A_p, cm^4	MLT, cm	N	AWG	$\Omega(50°C)$	P_Σ	$I = \sqrt{\dfrac{W}{\Omega}}$
17	52055	116.0	6.81	6.76	737		1.85	3.48	0.970
18	52012	110.0	9.35	8.88	505	20	1.66	3.30	0.996
19	52017	179.0	12.5	7.51	698	20	0.97	5.37	1.66
20	52031	256.0	19.8	8.23	1114	17	1.70	7.68	1.50
21	52103	220.0	24.5	8.77	688	17	1.12	6.60	1.72
22	52128	304.0	39.4	9.49	1104	17	1.94	9.12	1.53
23	52022	256.0	49.1	11.3	688	17	1.44	7.68	1.63
24	52042	347.0	78.7	12.0	1104	17	2.45	10.4	1.45
25	52100	422.0	145.0	15.4	1089	17	3.11	12.7	1.43
26	52112	878.0	510.0	20.3	2871	17	10.8	26.3	1.1
27	52426	1014.0	813.0	22.2	2856	17	11.7	24.4	1.02

Copper loss = iron loss

Definitions for Table 2.7

Information given is listed by column as:

1. Manufacturer part number (Magnetics Inc.). Other manufacturers of tape-wound cores are Arnold Engineering and Magnetic Metals.
2. Surface area calculated from Figure 2.22
3. Area product-effective iron area × window area
4. Mean length turn
5. Total number of turns and wire size using a window utilization factor $K_u = 0.40$
6. Resistance of the wire at 50°C
7. Watts loss is based on Figure 7.2 for a ΔT of 25°C with a room ambient of 25°C surface dissipation times the transformer surface area; total loss is equal to $2P_{cu}$
8. Current calculated from columns 6 and 7

9	10	11	12	13	14		15	16
$\Delta T(25°C)$ $J = I/cm^2$	$\Omega(75°C)$	P_Σ	$I = \sqrt{\dfrac{W}{\Omega}}$	$\Delta T(50°C)$ $J = I/cm^2$	Weight		Volume, cm^3	A_c, cm^2
					fe	cu		
187	2.0	8.12	1.42	273	83.4	220.0	86.4	0.686
192	1.82	7.70	1.45	280	143.0	235.0	87.4	1.371
160	1.065	12.5	2.33	274	107.0	455.0	163.0	0.686
145	1.86	17.9	2.19	211	131.0	800.0	272.0	0.686
165	1.23	15.4	2.51	241	238.0	503.0	212.0	1.371
147	2.13	21.3	2.24	215	286.0	896.0	341.0	1.371
157	1.58	17.9	2.38	229	477.0	629.0	291.0	2.742
140	2.69	24.3	2.12	204	572.0	1109.0	453.0	2.742
138	3.41	29.5	2.08	200	1117.0	1342.0	633.0	5.142
106	11.8	61.5	1.61	155	2205.0	4895.0	1891.0	6.855
98.1	12.9	71.0	1.66	159	3814.0	5077.0	2299.0	10.968

9. Current density calculated from columns 5 and 8
10. Resistance of the wire at 75°C
11. Watts loss is based on Figure 7.2 for a ΔT of 50°C with a room ambient of 25°C surface dissipation times the transformer surface area; total loss is equal to $2P_{cu}$
12. Current calculated from columns 10 and 11
13. Current density calculated from columns 5 and 12
14. Effective core weight plus copper, in grams
15. Transformer volume calculated from Figure 2.6
16. Core effective cross section

2.6 TRANSFORMER VOLUME AND A_p

The volume of a transformer can be related to the area product A_p of a transformer, treating the volume as shown in Figures 2.6–2.9 as a solid quantity without any subtraction for the core window. The relationship is derived according to the following reasoning: Volume varies in accordance with the cube of any linear dimension l whereas area product A_p varies as the fourth power:

$$\text{Volume} = K_1 l^3 \tag{2.1}$$

$$A_p = K_2 l^4 \tag{2.2}$$

$$l^4 = \frac{A_p}{K_2} \tag{2.3}$$

$$l = \left(\frac{A_p}{K_2}\right)^{0.25} \tag{2.4}$$

$$l^3 = \left[\left(\frac{A_p}{K_2}\right)^{0.25}\right]^3 = \left(\frac{A_p}{K_2}\right)^{0.75} \tag{2.5}$$

$$\text{Volume} = K_1\left(\frac{A_p}{K_2}\right)^{0.75} \tag{2.6}$$

$$K_v = \frac{K_1}{K_2^{0.75}} \tag{2.7}$$

The volume–area product relationship is therefore

$$\text{Volume} = K_v A_p^{0.75} \tag{2.8}$$

in which K_v is a constant related to core configuration whose values are given in Table 2.8. These values were obtained by averaging the values in column 15 of Tables 2.2–2.7.

The relationship between volume and area product A_p for various core types is graphed in Figures 2.10–2.15 from data presented in Tables 2.2–2.7, columns 3 and 15.

Table 2.8.   Constant K_v

Core type	K_v
Pot core	14.5
Powder core	13.1
Lamination	19.7
C core	17.9
Single-coil C core	25.6
Tape-wound core	25.0

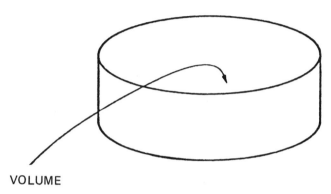

VOLUME

Figure 2.6.   Tape-wound core, powder core, and pot core volume.

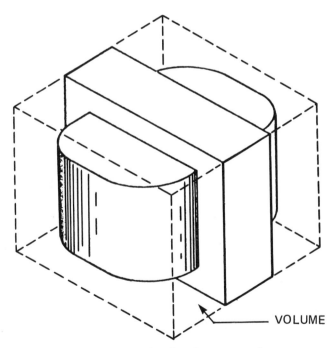

VOLUME

Figure 2.7.   EI lamination core volume.

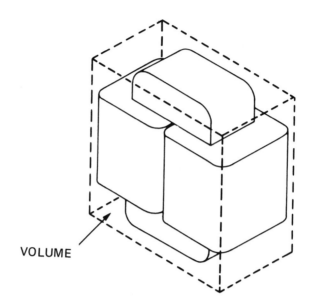

Figure 2.8. C core volume.

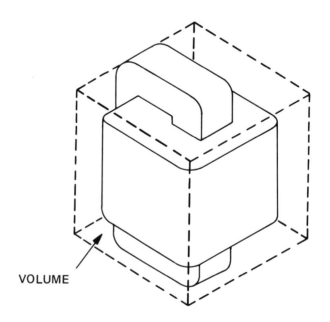

Figure 2.9. Single-coil C core volume.

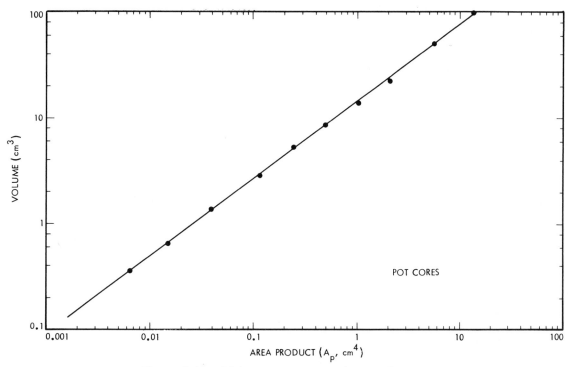

Figure 2.10.   Volume vs. area product A_p for pot cores.

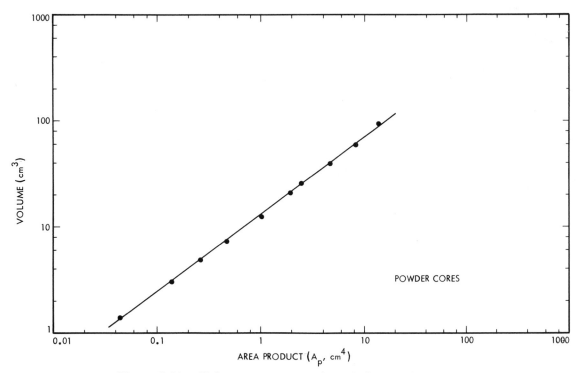

Figure 2.11.   Volume vs. area product A_p for powder cores.

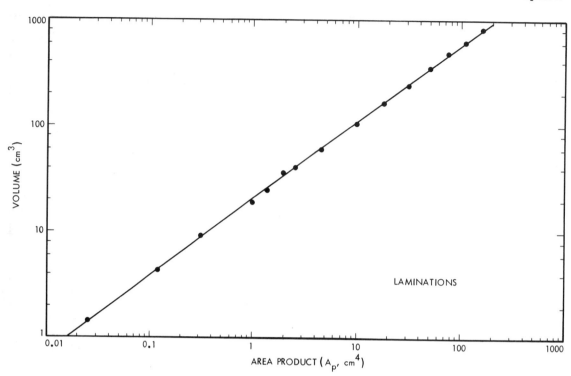

Figure 2.12.   Volume vs. area product A_p for laminations.

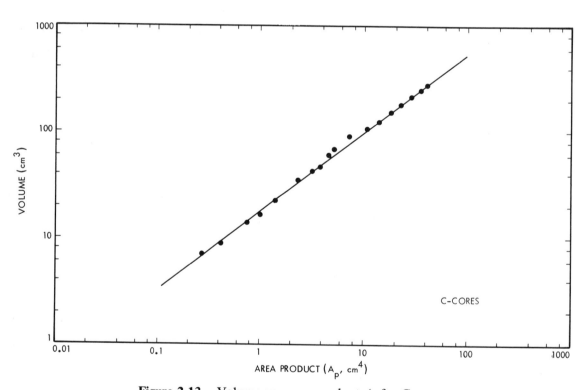

Figure 2.13.   Volume vs. area product A_p for C cores.

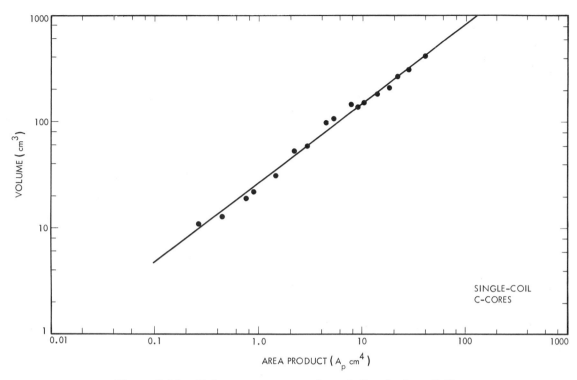

Figure 2.14.   Volume vs. area product A_p for single-coil C cores.

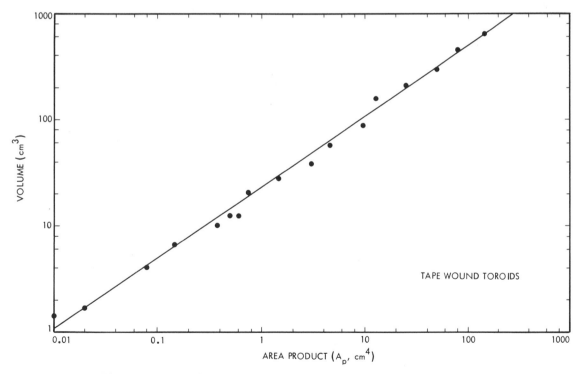

Figure 2.15.   Volume vs. area product A_p for tape-wound toroids.

2.7 TRANSFORMER WEIGHT AND A_p

The total weight W_t of a transformer can also be related to the area product A_p. The relationship is derived according to the following reasoning: Weight W_t varies in accordance with the cube of any linear dimension l, whereas area product A_p varies as the fourth power:

$$W_t = K_3 l^3 \tag{2.9}$$

$$A_p = K_2 l^4 \tag{2.10}$$

$$l^4 = \frac{A_p}{K_2} \tag{2.11}$$

$$l = \left(\frac{A_p}{K_2}\right)^{0.25} \tag{2.12}$$

$$l^3 = \left[\left(\frac{A_p}{K_2}\right)^{0.25}\right]^3 = \left(\frac{A_p}{K_2}\right)^{0.75} \tag{2.13}$$

$$W_t = K_3 \left(\frac{A_p}{K_2}\right)^{0.75} \tag{2.14}$$

$$K_w = \frac{K_3}{K_2^{0.75}} \tag{2.15}$$

The weight–area product relationship is therefore

$$W_t = K_w A_p^{0.75} \tag{2.16}$$

in which K_w is a constant related to core configuration whose values are given in Table 2.9. These values have been obtained by averaging the values in Tables 2.2–2.7, column 14.

The relationship between weight and area product A_p for various core types is graphed in Figures 2.16–2.21 from the data presented in Tables 2.2–2.7, columns 3 and 14.

Table 2.9. Constant K_w

Core type	K_w
Pot core	48.0
Powder core	58.8
Lamination	68.2
C core	66.6
Single-coil C core	76.6
Tape-wound core	82.3

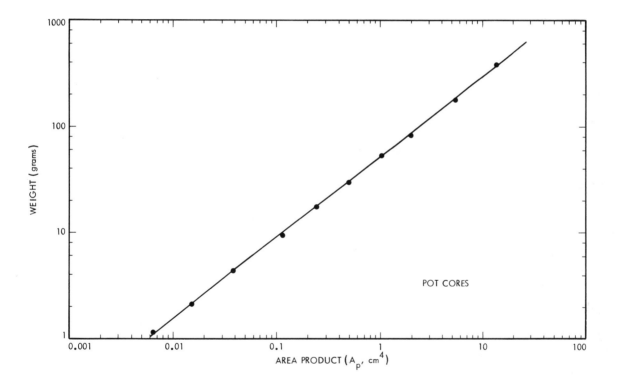

Figure 2.16. Total weight vs. area product A_p for pot cores.

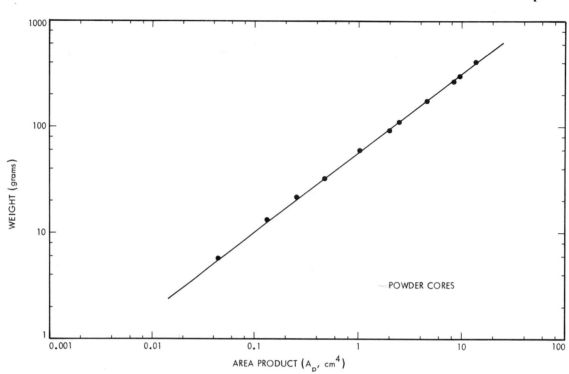

Figure 2.17. Total weight vs. area product A_p for powder cores.

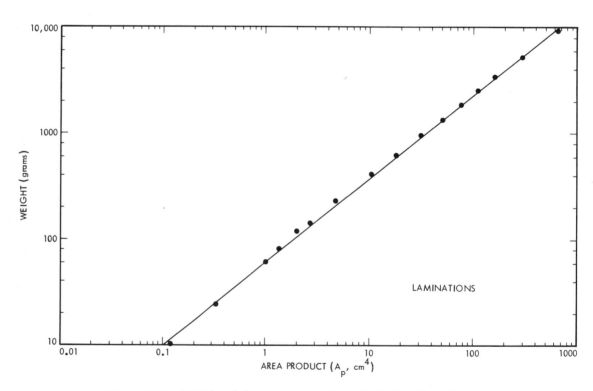

Figure 2.18. Total weight vs. area product A_p for laminated cores.

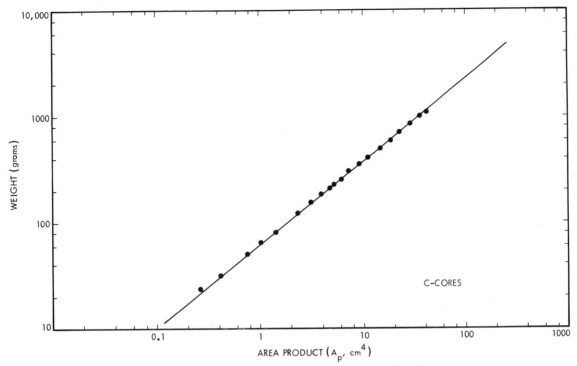

Figure 2.19. Total weight vs. area product A_p for C cores.

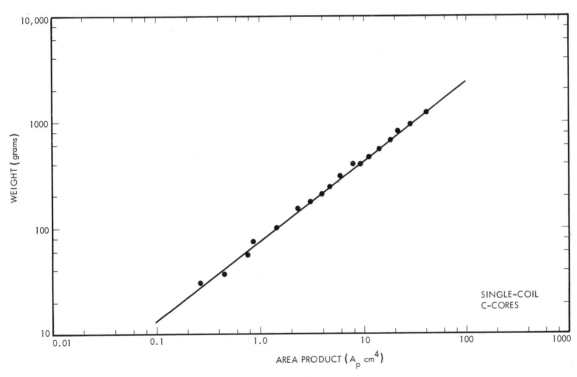

Figure 2.20. Total weight vs. area product A_p for single-coil C cores.

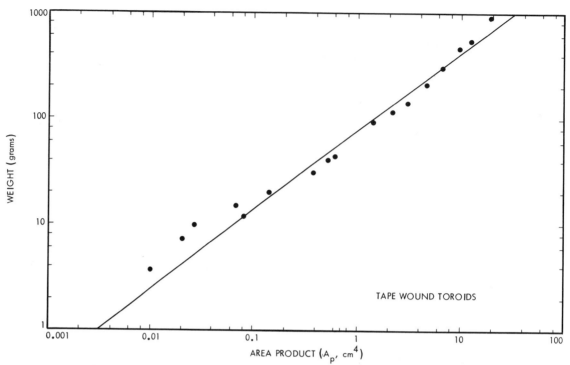

Figure 2.21. Total weight vs. area product A_p for tape-wound toroids.

2.8 TRANSFORMER SURFACE AREA AND A_p

The surface area A_t of a transformer can be related to the area product A_p of a transformer by treating the surface area as shown in Figures 2.22–2.25. The relationship is derived in accordance with the following reasoning: The surface area varies with the square of any linear dimension l, whereas the area product varies as the fourth power:

$$A_t = K_4 l^2 \tag{2.17}$$

$$A_p = K_2 l^4 \tag{2.18}$$

$$l^4 = \frac{A_p}{K_2} \tag{2.19}$$

$$l = \left(\frac{A_p}{K_2}\right)^{0.25} \tag{2.20}$$

$$l^2 = \left[\left(\frac{A_p}{K_2}\right)^{0.25}\right]^2 \tag{2.21}$$

$$l^2 = \left(\frac{A_p}{K_2}\right)^{0.5} \tag{2.22}$$

$$A_t = K_4\left(\frac{A_p}{K_2}\right)^{0.5} \tag{2.23}$$

$$K_s = \frac{K_4}{K_2^{0.5}} \tag{2.24}$$

The relationship between surface area and area product can therefore be expressed as

$$A_t = K_s A_p^{0.5} \tag{2.25}$$

in which K_s is a constant related to core configuration whose values are given in Table 2.10. These values have been derived by averaging the values in Tables 2.2–2.7, column 2.

The relationship between surface area and area product A_p for various core types is graphed in Figures 2.26–2.31 from the data of Tables 2.2–2.7, columns 2 and 3.

Table 2.10. Constant K_s

Core type	K_s
Pot core	33.8
Powder core	32.5
Lamination	41.3
C core	39.2
Single-coil C core	44.5
Tape-wound core	50.9

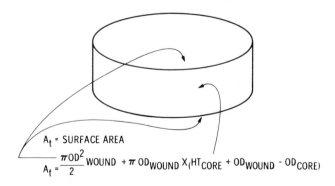

A_t = SURFACE AREA

$$A_t = \frac{\pi OD^2}{2} WOUND + \pi OD_{WOUND} \times (HT_{CORE} + OD_{WOUND} - OD_{CORE})$$

Figure 2.22. Surface area A_t for a tape-wound core, powder core, or pot core.

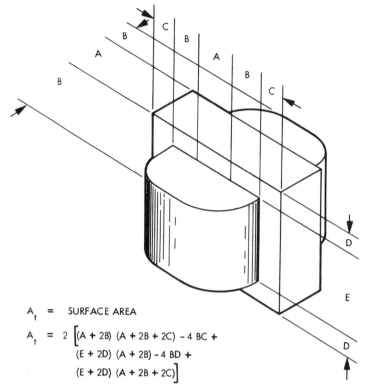

$$A_t \; = \; \text{SURFACE AREA}$$

$$A_t \; = \; 2 \left[(A + 2B)\,(A + 2B + 2C) - 4\,BC \; + \right.$$
$$(E + 2D)\,(A + 2B) - 4\,BD \; +$$
$$\left. (E + 2D)\,(A + 2B + 2C) \right]$$

Figure 2.23.   Surface area A_t for an EI lamination.

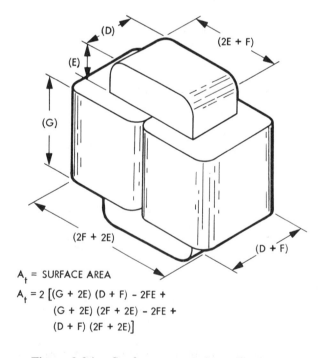

$$A_t = \text{SURFACE AREA}$$

$$A_t = 2 \left[(G + 2E)\,(D + F) - 2FE \; + \right.$$
$$(G + 2E)\,(2F + 2E) - 2FE \; +$$
$$\left. (D + F)\,(2F + 2E) \right]$$

Figure 2.24.   Surface area A_t for a C core.

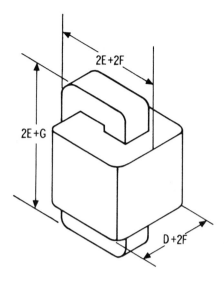

$$A_t = 2\left\{\ 2(E+F)\left[(D+2F)+(G+2E)\right]+(G+2E)(D+2F)-8EF\right\}$$

Figure 2.25. Surface area A_t for a single-coil C core.

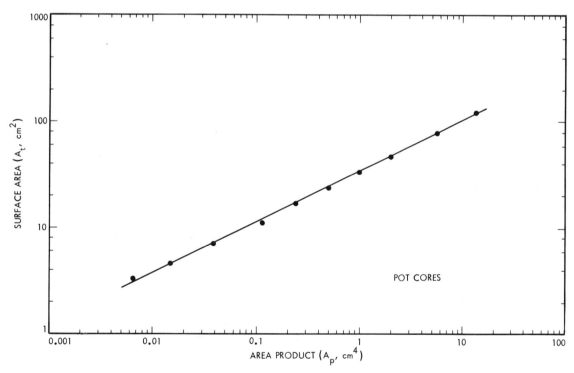

Figure 2.26. Surface area vs. area product A_p for pot cores.

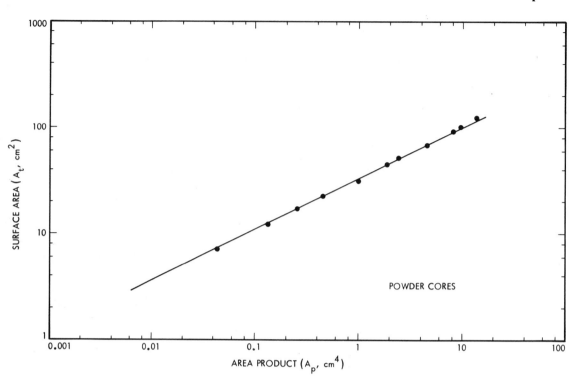

Figure 2.27.   Surface area vs. area product A_p for powder cores.

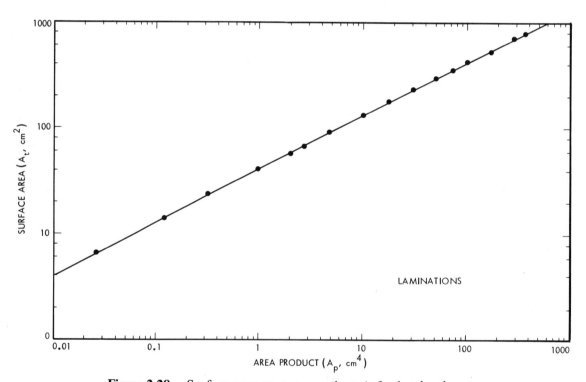

Figure 2.28.   Surface area vs. area product A_p for laminations.

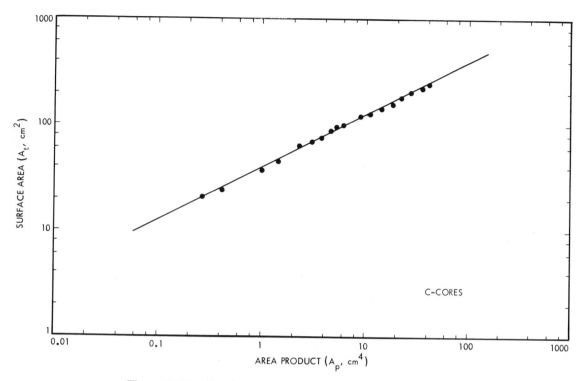

Figure 2.29. Surface area vs. area product A_p for C cores.

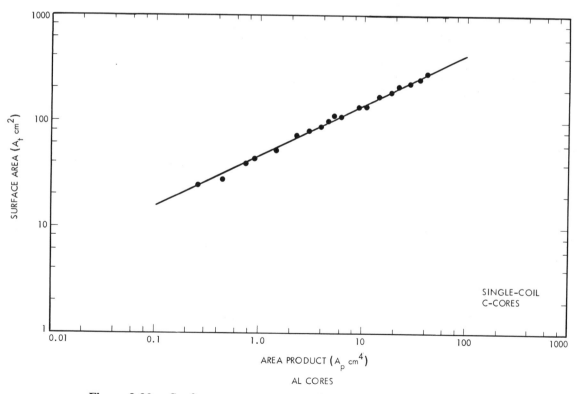

Figure 2.30. Surface area vs. area product A_p for single-coil C cores.

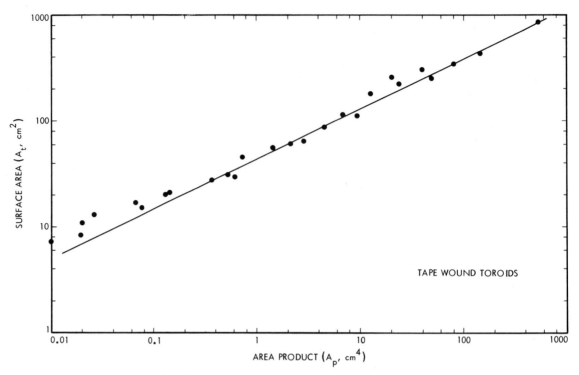

Figure 2.31. Surface area versus area product A_p for tape-wound toroids.

2.9 TRANSFORMER CURRENT DENSITY AND A_p

Current density J of a transformer can be related to the area product A_p of a transformer for a given temperature rise. The relationship can be derived as follows:

$$A_t = K_s A_p^{0.5} \qquad (2.26)$$

$$P_{cu} = I^2 R \qquad (2.27)$$

$$I = A_w J \qquad (2.28)$$

Therefore,

$$P_{cu} = A_w^2 J^2 R \qquad (2.29)$$

and since

$$R = \frac{\text{MLT}}{A_w} N\rho \qquad (2.30)$$

we have

$$P_{cu} = A_w^2 J^2 \frac{MLT}{A_w} N\rho \tag{2.31}$$

$$P_{cu} = A_w J^2 \, MLT \, N\rho \tag{2.32}$$

Since MLT has a dimension of length,

$$MLT = K_5 A_p^{0.25} \tag{2.33}$$

$$P_{cu} = A_w J^2 K_5 A_p^{0.25} N\rho \tag{2.34}$$

$$A_w N = K_6 W_a = K_3 A_p^{0.5} \tag{2.35}$$

$$P_{cu} = K_6 A_p^{0.5} K_5 A_p^{0.25} J^2 \rho \tag{2.36}$$

Let

$$K_7 = K_6 K_5 \rho \tag{2.37}$$

Then, assuming the core loss is the same as the copper loss for optimized transformer operation (see Chapter 7),

$$P_{cu} = K_7 A_p^{0.75} J^2 = P_{fe} \tag{2.38}$$

$$P_\Sigma = P_{cu} + P_{fe} \tag{2.39}$$

$$\Delta T = K_8 \frac{P_\Sigma}{A_t} \tag{2.40}$$

$$\Delta T = \frac{2 K_8 K_7 J^2 A_p^{0.75}}{K_s A_p^{0.5}} \tag{2.41}$$

To simplify, let

$$K_9 = \frac{2 K_8 K_7}{K_s} \tag{2.42}$$

Then

$$\Delta T = K_9 J^2 A_p^{0.25} \tag{2.43}$$

$$J^2 = \frac{\Delta T}{K_9 A_p^{0.25}} \tag{2.44}$$

Then, letting

$$K_{10} = \frac{\Delta T}{K_9} \tag{2.45}$$

we have

$$J^2 = K_{10} A_p^{-0.25} \tag{2.46}$$

The relationship between current density and area product can therefore be expressed as*

$$J = K_j A_p^{-0.125} \tag{2.47}$$

in which K_j is a constant related to core configuration whose values are given in Table 2.11. These values have been derived by averaging the values in Tables 2.2–2.7, columns 9 and 13.

The relationship between current density and area product A_p for temperature increases of 25°C and 50°C is graphed in Figs. 2.32–2.37 from the data of Tables 2.2–2.7, columns 9 and 13.

Table 2.11. Constant K_j for temperature increases of 25°C and 50°C

Core type	K_j ($\Delta 25°$)	K_j ($\Delta 50°$)
Pot core	433	632
Powder core	403	590
Lamination	366	534
C-type core	322	468
Single-coil C core	395	569
Tape-wound core	250	365

---

*This is the theoretical relationship. The empirical values for various core configurations are given in Table 2.1.

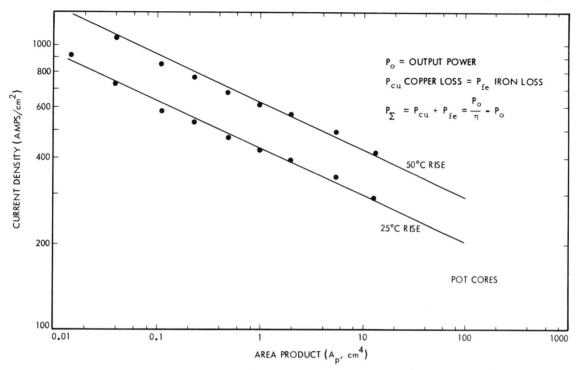

Figure 2.32. Current density vs. area product A_p for temperature increases of 25°C and 50°C for pot cores.

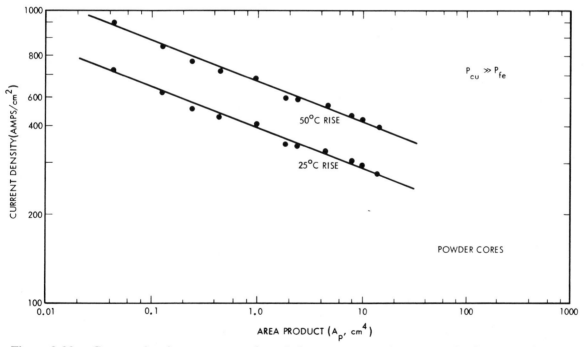

Figure 2.33. Current density vs. area product A_p for temperature increases of 25°C and 50°C for powder cores.

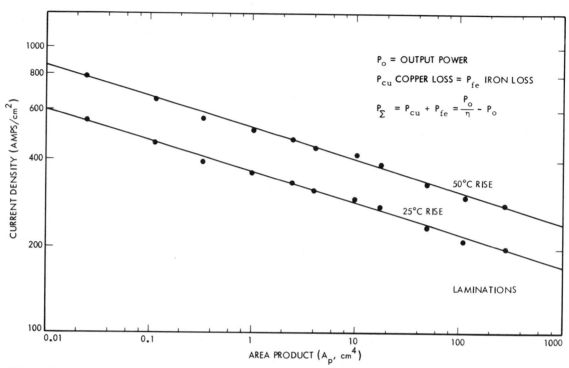

Figure 2.34. Current density vs. area product A_p for temperature increases of 25°C and 50°C for laminations.

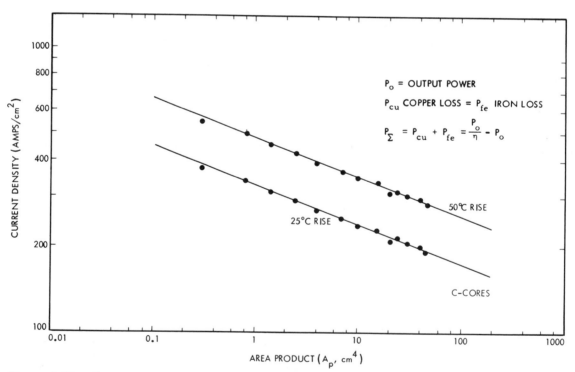

Figure 2.35. Current density vs. area product A_p for temperature increases of 25°C and 50°C for C cores.

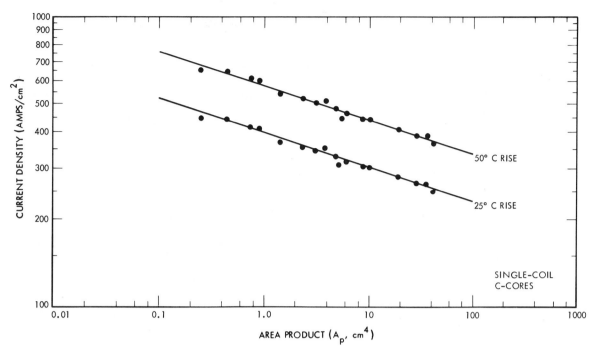

Figure 2.36. Current density vs. area product A_p for temperature increases of 25°C and 50°C for single-coil C cores.

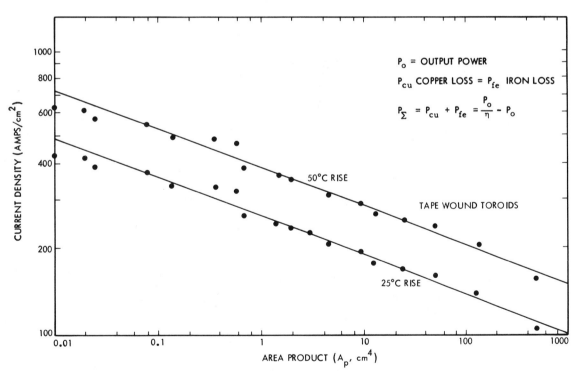

Figure 2.37. Current density vs. area product A_p for temperature increases of 25°C and 50°C for tape-wound toroids.

2.10 TRANSFORMER AREA PRODUCT AND GEOMETRY

The geometry K_g of a transformer, which can be related to the area product A_p, is derived in Chapter 3 and is shown here in Equation (2.48). Derivation of the relationship is according to the following: Geometry K_g varies in accordance with the fifth power of any linear dimension l, whereas area product A_p varies as the fourth power:

$$K_g = \frac{W_a A_c^2 K_u}{\text{MLT}} \tag{2.48}$$

$$K_g = K_{10} l^5 \tag{2.49}$$

$$A_p = K_2 l^4 \tag{2.50}$$

From Eq. (2.49),

$$l = \left(\frac{K_g}{K_{10}}\right)^{0.20} \tag{2.51}$$

Then

$$l^4 = \left[\left(\frac{K_g}{K_{10}}\right)^{0.20}\right]^4 = \left(\frac{K_g}{K_{10}}\right)^{0.8} \tag{2.52}$$

Substituting (2.52) into (2.50),

$$A_p = K_2 \left(\frac{K_g}{K_{10}}\right)^{0.8} \tag{2.53}$$

Let

$$K_p = \frac{K_2}{K_{10}^{0.8}} \tag{2.54}$$

Then

$$A_p = K_p K_g^{0.8} \tag{2.55}$$

where K_p is a constant related to core configuration whose values are given in Table 2.12. These values have been derived by averaging the values in Tables 2.2–2.7 and Tables 3.3–3.10.

The relationship between are product A_p and core geometry is graphed in Figures 2.38–2.42 from the data shown in Tables 2.2–2.7 for area product A_p and Tables 3.3–3.10 for K_g.

Table 2.12. Constant K_p

Core type	K_p
Pot core	8.87
Powder core	11.8
Lamination	8.3
C core	12.5
Tape-wound core	

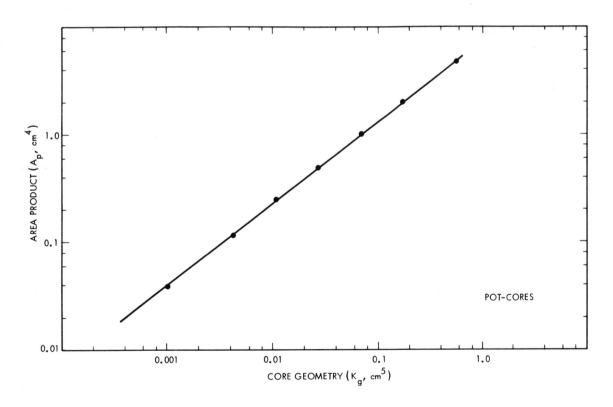

Figure 2.38. Area product vs. core geometry for pot cores.

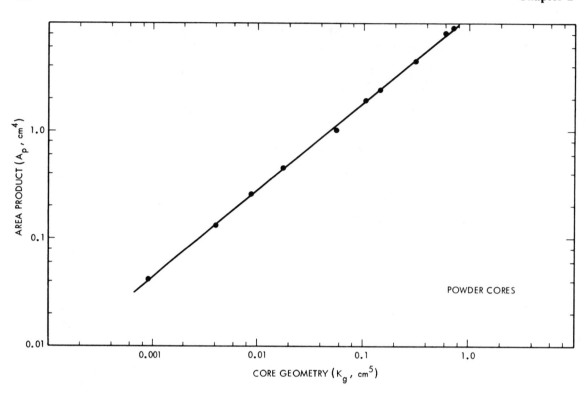

Figure 2.39. Area product vs. core geometry for powder cores.

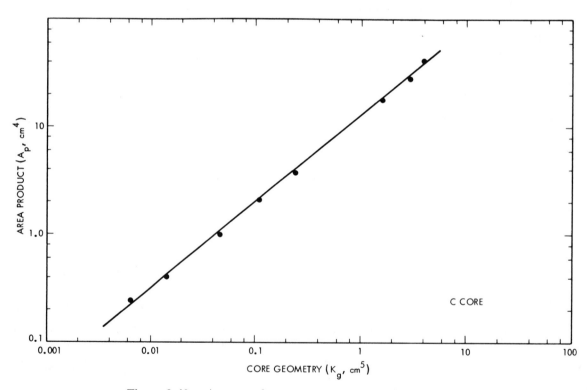

Figure 2.40. Area product vs. core geometry for C cores.

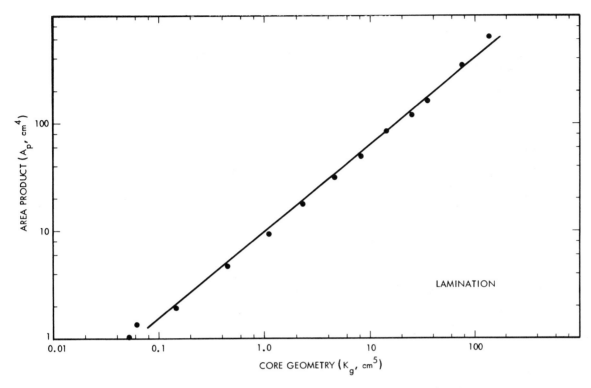

Figure 2.41.   Area product vs. core geometry for laminations.

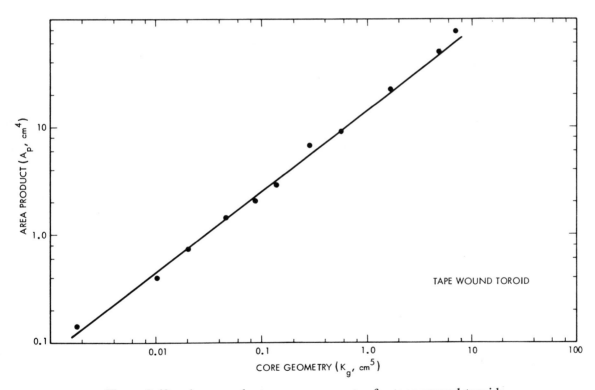

Figure 2.42.   Area product vs. core geometry for tape-wound toroids.

BIBLIOGRAPHY

1. C. McLyman, *Transformer Design Tradeoffs,* Technical Memorandum 33-767 Rev. 1, Jet Propulsion Laboratory, Pasadena, Calif.

2. W. J. Muldoon, *High Frequency Transformer Optimization,* HAC Trade Study Report 2228/1130, May, 1970.

3. R. G. Klimo, A. B. Larsen, and J. E. Murray, *Optimization Study of High Power Static Inverters and Converters,* Quarterly Report No. 2 NASA-CR-54021, April 20, 1964, Contract NAS 3-2785.

4. F. F. Judd and D. R. Kressler, *Design Optimization of Power Transformers,* Bell Laboratories, Whippany, New Jersey IEEE Applied Magnetics Workshop, June 5–6, 1975.

Chapter 3

Power Transformer Design

3.1 INTRODUCTION

The conversion process in power electronics requires the use of transformers, components that are frequently the heaviest and bulkiest item in the conversion circuits. They also have a significant effect upon the overall performance and efficiency of the system. Accordingly, the design of such transformers has an important influence on overall system weight, power conversion efficiency, and cost. Because of the interdependence and interaction of parameters, judicious tradeoffs are necessary to achieve design optimization.

3.2 THE DESIGN PROBLEM GENERALLY

The designer is faced with a set of constraints that must be observed in the design of any transformer. One of these is the output power P_o (operating voltage multiplied by maximum current demand) that the secondary winding must be capable of delivering to the load within specified regulation limits. Another relates to minimum efficiency of operation, which is dependent upon the maximum power loss that can be allowed in the transformer. Still another defines the maximum permissible temperature rise for the transformer when it is used in a specified temperature environment.

One of the basic steps in transformer design is the selection of the proper core material. To aid in the selection of cores, a comparison of five common core materials is presented here to illustrate their influence on overall transformer efficiency and weight. The designer should also be aware of the cost difference between core materials of the nickel-steel and silicon-steel families. In many instances, the author has found it possible to achieve suitable designs using low-cost silicon steel C cores when the proper design tradeoffs were made.

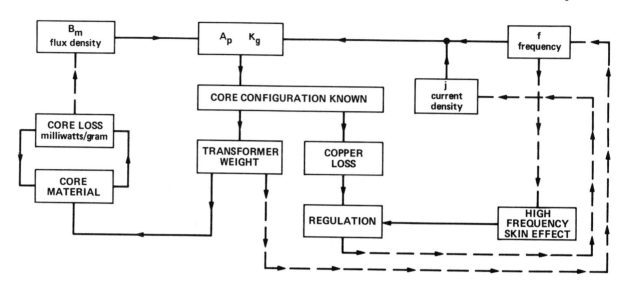

Figure 3.1. Transformer design factors flow chart.

Other constraints relate to volume occupied by the transformer and, particularly in aerospace applications, weight, since weight minimization is an important goal in the design of space flight electronics. Finally, cost effectiveness is always an important consideration.

Depending upon the application, certain of these constraints will dominate. Parameters affecting others may then be traded off as necessary to achieve the most desirable design. It is not possible to optimize all the parameters in a single design because of their interaction and interdependence. For example, if volume and weight are of great significance, reductions in both can often be effected by operating the transformer at a higher frequency but at a penalty in efficiency. When the frequency cannot be increased, reduction in weight and volume may still be possible by selecting a more efficient core material, but at the penalty of increased cost. Judicious tradeoffs thus must be effected to achieve the design goals.

A flow chart showing the interrelation and interaction of the various design factors that must be taken into consideration is shown in Figure 3.1.

Transformer designers have used various approaches in arriving at suitable designs. For example, in many cases a rule of thumb is used for dealing with current density. Typically, an assumption is made that a good working level is 1000 circular mils per ampere. This will work in many instances, but the wire size needed to meet this requirement may produce a heavier and bulkier transformer than desired or required. The information presented in this volume makes it possible to avoid the use of this and other rules of thumb and to develop a more economical design with great accuracy.

3.3 POWER-HANDLING ABILITY

Manufacturers have for years assigned numeric codes to their cores; these codes represent the power-handling ability. This method assigns to each core a number that is the product of its window area W_a and core cross section area A_c and is called area product A_p.

These numbers are used by core suppliers to summarize dimensional and electrical properties in their catalogs. They are available for laminations, C-cores, pot cores, powder cores, ferrite toroids, and toroidal tape-wound cores.

The regulation and power-handling ability of a core is related to the core geometry K_g. Every core has its own inherent K_g. The core geometry is relatively new, and magnetic core manufacturers do not list this coefficient.

Because of their significance, the area product A_p and core geometry K_g are treated extensively. A great deal of other information is also presented for the convenience of the designer. Much of the material is in tabular form to assist the designer in making the tradeoffs best suited for his particular application in a minimum amount of time.

These relationships can now be used as new tools to simplify and standardize the process of transformer design. They make it possible to design transformers of lighter weight and smaller volume or to optimize efficiency without going through a cut and try design procedure. While developed specifically for aerospace applications, the information has wider utility and can be used for the design of now-aerospace transformers as well.

3.4 OUTPUT POWER VERSUS APPARENT POWER CAPABILITY

Output power P_o is of greatest interest to the user. To the transformer designer, the apparent power P_t, which is associated with the geometry of the transformer, is of greater importance. Assume, for the sake of simplicity, that the core of an isolation transformer has only two windings in the window area, a primary and a secondary. Also assume that the window area W_a is divided up in proportion of the power-handling capability of the windings using equal current density. The primary winding handles P_{in}, and the secondary handles P_o to the load. Since the power transformer has to be designed to accommodate the primary P_{in} and P_o, then by definition,

$$P_t = P_{in} + P_o \quad \text{[watts]}$$

$$P_{in} = \frac{P_o}{\eta} \quad \text{[watts]}$$

The primary turns can be expressed using Faraday's law:

$$N_p = \frac{V_p \times 10^4}{K_f B_m A_c f} \quad \text{[turns]}$$

The winding area of a transformer is fully utilized when

$$K_u W_a = N_p A_{wp} + N_s A_{ws}$$

where K_u is the window utilization factor (see Chapter 6)
W_a is the window area, cm^2
N is the number of turns
A_w is the wire area, cm^2

By definition,

$$A_w = \frac{I}{J}$$

where I is the rms current
J is the current density, A/cm^2

Rearranging the equation:

$$K_u W_a = N_p \frac{I_p}{J} + N_s \frac{I_s}{J}$$

Substituting into Faraday's equation, we get

$$K_u W_a = \left(\frac{V_p \times 10^4}{K_f B_m A_c f} \right) \left(\frac{I_p}{J} \right) + \left(\frac{V_s \times 10^4}{K_f B_m A_c f} \right) \left(\frac{I_s}{J} \right)$$

Rearranging:

$$W_a A_c = \frac{(V_p I_p + V_s I_s) \times 10^4}{K_u K_f B_m f J}$$

The output power P_o is

$$P_o = V_s I_s$$

and the input power P_{in} is

$$P_{\text{in}} = V_p I_p$$

Then

$$P_t = P_{\text{in}} + P_o$$

Therefore

$$W_a A_c = \frac{P_t \times 10^4}{K_u K_f B_m f J}$$

By definition,

$$A_p = W_a A_c$$

so

$$A_p = \frac{P_t \times 10^4}{K_u K_f B_m f J}$$

The designer must be concerned with the apparent power-handling capability P_t of the transformer core and windings. P_t may vary by a factor ranging from 2 to 2.828 times the input power P_{in}, depending upon the type of circuit in which the transformer is used. If the current in the rectifier transformer becomes interrupted, its effective rms value changes. Transformer size is thus determined not only by the load demand but also by application because of the different copper losses incurred owing to current waveform.

For example, for a load of 1 W, compare the power-handling capabilities required for each winding (neglecting transformer and diode losses so that $P_{\text{in}} = P_o$) for the full-wave bridge circuit of Figure 3.2 the full-wave center-tapped secondary circuit of Figure 3.3, and the push-pull center-tapped full-wave circuit in Figure 3.4, where all windings have the same number of turns, N.

The total apparent power P_t for the circuit shown in Figure 3.2 is 2 W. This is shown in the following equation:

$$P_t = 2P_{\text{in}} \quad [\text{watts}]$$

The total power P_t for the circuit shown in Figure 3.3 increased 20.7% owing to the distorted wave form of the interrupted current flowing in the secondary winding. This is shown in the equation

$$P_t = P_o\left(\frac{1}{\eta} + \sqrt{2}\right) \quad [\text{watts}]$$

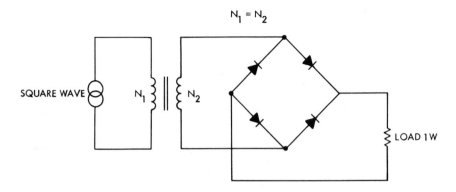

Figure 3.2.   Full-wave bridge circuit.

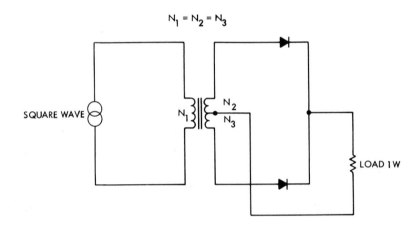

Figure 3.3.   Full-wave center-tapped circuit.

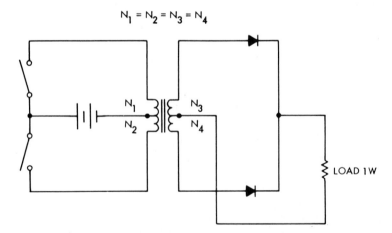

Figure 3.4.   Push-pull, full-wave center-tapped circuit.

The total power P_t for the circuit shown in Figure 3.4, which is typical of a dc-to-dc converter, increases to 2.828 P_{in} because of the interrupted current flowing in both the primary and secondary windings, since

$$P_t = P_o\left(\frac{\sqrt{2}}{\eta} + \sqrt{2}\right) \quad \text{[watts]}$$

3.4.1 Transformers with Multiple Outputs

The following is an example of how the apparent power P_t changes with a multiple-output transformer.

Output	Circuit
5 V at 10 A	Center-tapped
15 V at 1 A	Full-wave bridge
V_d = diode drop = 1 V	
η = 0.95	

The output power seen by the transformer in Figure 3.5 is

$$P_{o1} = (V_{o1} + V_d)(I_o)$$

$$P_{o1} = (5 + 1)(10)$$

$$P_{o1} = 60 \quad \text{[W]}$$

and

$$P_{o2} = (V_{o2} + 2V_d)(I_{o2})$$

$$P_{o2} = (15 + 2)(1)$$

$$P_{o2} = 17 \quad \text{[W]}$$

Because of the different winding configurations, the apparent power P_t of the transformer will have to be summed to reflect this. When a winding has a center tap and produces a discontinuous current, then the power in that winding, whether it is primary or secondary, has to be multiplied by

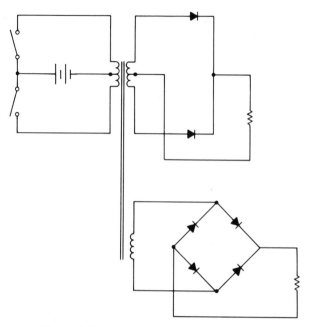

Figure 3.5. Multiple-output converter.

the factor U to correct for the rms current in that winding. If the winding has a center tap, then $U = 1.41$; if not, the $U = 1$.

For example, summing the output power of a multiple-output transformer would be:

$$P_\Sigma = P_{o1}(U) + P_{o2}(U) + \cdots + P_{on}(U)$$

Then for our example,

$$P_\Sigma = (P_{o1}U) + (P_{o2}U)$$

$$P_\Sigma = (60)(1.41) + (17)(1)$$

$$P_\Sigma = 101.6$$

After the secondary has been totaled, the primary power can be calculated:

$$\text{Primary:} \quad \frac{P_\Sigma}{\eta} = P_{\text{in}}$$

$$P_{\text{in}} = \frac{P_{\Sigma}}{\eta}$$

$$P_{\text{in}} = \frac{101.6}{0.95}$$

$$P_{\text{in}} = 106.3$$

Then the apparent power P_t is

$$P_t = P_{\text{in}}(U) + P_{\Sigma} \quad \text{[watts]}$$

$$P_t = (106.3)(1.41) + 101.6$$

$$P_t = 251 \quad \text{[W]}$$

3.5 REGULATION

The minimum size of a transformer is usually determined either by a temperature rise limit or by allowable voltage regulation, assuming that size and weight are to be minimized. Regulation is denoted by α (Greek alpha) and is expressed in percent.

Figure 3.6 shows the circuit diagram of a transformer with one secondary.

This assumes that distributed capacitance in the secondary can be neglected because the frequency and secondary voltage are not excessively high. Also, the winding geometry is designed to limit the leakage inductance to a level low enough to be neglected under most operating conditions.

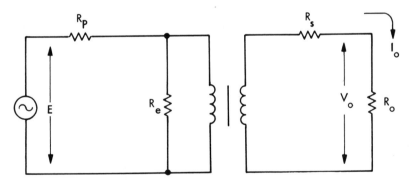

Figure 3.6.   Transformer circuit diagram.

Transformer voltage regulation can now be expressed as

$$\alpha = \frac{V_o(\text{NL}) - V_o(\text{FL})}{V_o(\text{FL})} \times 100 \ [\%]$$

in which $V_o(\text{NL})$ is the no-load voltage and $V_o(\text{FL})$ is the full-load voltage.

For the sake of simplicity, assume that the transformer in Figure 3.6 is an isolation transformer with a 1:1 turns ratio and that the core impedance R_e is infinite.

If the transformer has a 1:1 turns ratio and the core impedance is infinite, then

$$I_{\text{in}} = I_o \quad \text{and} \quad R_p = R_s$$

With equal window area allocated for the primary and secondary windings and using the same current density J,

$$\Delta V_p = I_{\text{in}} R_p = \Delta V_s = I_o R_s$$

Regulation is then

$$\alpha = \frac{\Delta V_p}{V_p} \times 100 + \frac{\Delta V_s}{V_s} \times 100$$

Multiplying the equation by the current I,

$$\alpha = \left(\frac{\Delta V_p I_{\text{in}}}{V_p I_{\text{in}}} \times 100 \right) + \left(\frac{\Delta V_s I_o}{V_s I_o} \times 100 \right)$$

Primary copper loss is

$$P_p = \Delta V_p I_{\text{in}}$$

Secondary copper loss is

$$P_s = \Delta V_s I_o$$

Total copper loss is

$$P_{cu} = P_p + P_s$$

Then the regulation equation can be rewritten to

$$\alpha = \frac{P_{cu}}{P_o} \times 100$$

Regulation can be expressed as the power lost in the copper. A transformer with an output power of 100 W and regulation of 2% will have a 2-W loss in the copper:

$$P_{cu} = \frac{P_o}{100} \quad \text{[watts]}$$

$$P_{cu} = \frac{(100)(2)}{100}$$

$$P_{cu} = 2 \quad \text{[W]}$$

3.5.1 Relationship of K_g to Power Transformer Regulation Capability

Although most transformers are designed for a given temperature rise, they can also be designed for a given regulation. The regulation and power-handling ability of a core are related to two constants, K_g and K_e, by the following equation:

$$\alpha = \frac{P_t}{2K_g K_e} \quad \text{[%]}$$

where α is the regulation, %.

The constant K_g is the *core geometry coefficient* and is determined by the core geometry according to the equation

$$K_g = \frac{W_a A_c^2 K_u}{\text{MLT}} \quad \text{[cm}^5\text{]}$$

The constant K_e is determined by the magnetic and electric operating conditions, which may be related by the equation

$$K_e = 0.145 K_f^2 f^2 B_m^2 \times 10^{-4}$$

where K_f is the waveform coefficient,

$$K_f = \begin{cases} 4.44 & \text{for a sine wave} \\ 4.0 & \text{for a square wave} \end{cases}$$

3.5.2 Relationship of A_p to Transformer Power-Handling Capability

According to the newly developed approach, the power-handling capability of a core is related to its area product by an equation that may be stated as:

$$A_p = \left(\frac{P_t \times 10^4}{K_f B_m f K_u K_j} \right)^{(x)} \quad [\text{cm}^4]$$

where the value of the exponent (x) is given in Table 3.1 and K_f is the waveform coefficient

$$K_f = \begin{cases} 4.44 & \text{for a sine wave} \\ 4.0 & \text{for a square wave} \end{cases}$$

From the above it can be seen that factors such as flux density; frequency of operation f; the window utilization factor K_u, which defines the maximum space which may be occupied by the copper in the window (see Chapter 6); and the constant K_j, which is related to temperature rise, all have an influence on the transformer area product. The constant K_j is a new parameter that gives the designer control of the copper loss.

Table 3.1. Core configuration constants*

Core	Losses	K_j (25°C)	K_j (50°C)	(x)	(y)
Pot core	$P_{cu} = P_{fe}$	433	632	1.20	−0.17
Powder core	$P_{cu} \gg P_{fe}$	403	590	1.14	−0.12
Lamination	$P_{cu} = P_{fe}$	366	534	1.14	−0.12
C core	$P_{cu} = P_{fe}$	323	468	1.16	−0.14
Single-coil	$P_{cu} \gg P_{fe}$	395	569	1.16	−0.14
Tape-wound core	$P_{cu} = P_{fe}$	250	365	1.15	−0.13

*(x) and (y) are exponents in equations for A_p and J.

$$A_p = \left(\frac{P_t \times 10^4}{K_f B_m f K_u K_j} \right)^{(x)} \quad [\text{cm}^4]$$

3.6 THE AREA PRODUCT A_p AND ITS RELATIONSHIPS

The A_p of a core is the product of the available window area W_a of the core in square centimeters (cm^2) multiplied by the effective cross-sectional area A_c in square centimeters (cm^2), which may be stated as

$$A_p = W_a A_c \quad [\text{cm}^4]$$

Figures 3.7–3.11 show in outline form five types of transformer cores that are typical of those shown in suppliers' catalogs.

There is a unique relationship between the area product A_p characteristic number for transformer cores and several other important parameters that must be considered in transformer design (see Table 3.1).

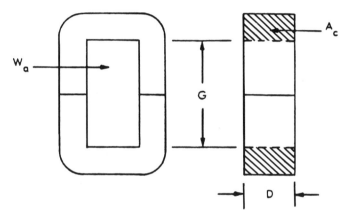

Figure 3.7. C-core.

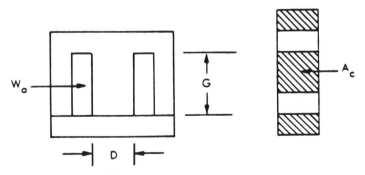

Figure 3.8. EI lamination.

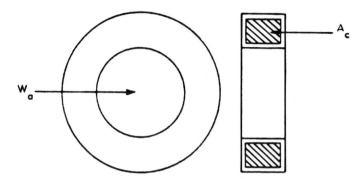

Figure 3.9. Tape-wound toroidal core.

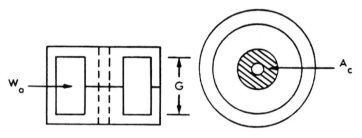

Figure 3.10. Pot core.

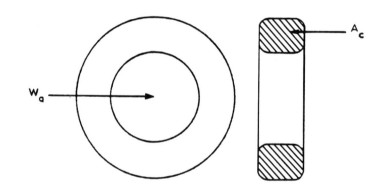

Figure 3.11. Powder core and ferrites.

3.7 MAGNETIC CORE MATERIAL TRADEOFF

The area product A_p and core geometry K_g are associated only with such geometric properties as surface area and volume, weight, and the factors affecting temperature rise such as current density. They have no relevance to the magnetic core materials used. However, the designer must often make tradeoffs between such goals as efficiency and size, which are influenced by core material selection.

Usually in articles written about inverter and converter transformer design, recommendations with respect to choice of core material are a compromise of material characteristics such as those tabulated in Table 3.2 and graphically displayed in Figure 3.12. The characteristics shown here are

Table 3.2. Magnetic core material characteristics

Trade name	Composition	Saturated flux density, tesla *	DC coercive force, AMP-TURN/ cm	Squareness ratio	Material density, g/cm³ **	Curie temperature, °C	Weight factor
Supermendur	49% Co 49% Fe 2% V	1.9–2.2	0.18–0.44	0.90–1.0	8.15	930	1.066
Permendur Magnesil Silectron Microsil Supersil	3% Si 97% Fe	1.5–1.8	0.5–0.75	0.85–0.75	7.63	750	1.00
Deltamax Orthonol 49 Sq Mu	50% Ni 50% Fe	1.4–1.6	0.125–0.25	0.94–1.0	8.24	500	1.079
Allegheny 4750 48 Alloy Carpenter 49	48% Ni 52% Fe	1.15–1.4	0.062–0.187	0.80–0.92	8.19	480	1.073
4-79 Permalloy Sp Permalloy 80 Sq Mu 79	79% Ni 17% Fe	0.66–0.82	0.025–0.82	0.80–1.0	8.73	460	1.144
Supermalloy	78% Ni 17% Fe 5% Mo	0.65–0.82	0.0037–0.01	0.40–0.70	8.76	400	1.148
Ferrites F N27 3C8	Mn Zn	0.45–0.50	0.25	0.30–0.5	4.8	250	0.629

*tesla = 10^4 Gauss
**g/cm³ = 0.036 lb/in³

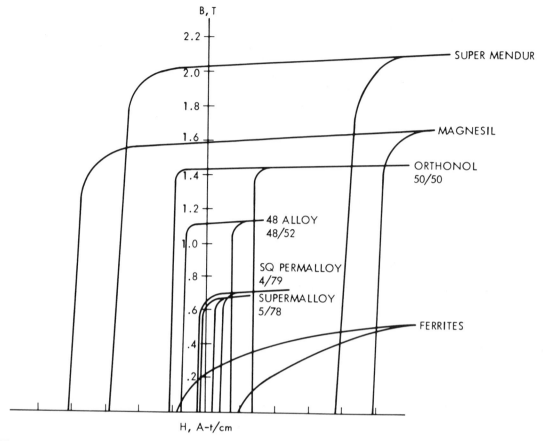

Figure 3.12. Typical dc *B-H* loops for some magnetic core materials (Courtesy of Magnetics).

those typical of commercially available core materials. As can be seen, the core material that provides the highest flux density is Supermendur. It also produces the smallest component size. If size is the most important consideration, this should determine the choice of materials. On the other hand, type 78 Supermalloy (see the 5/78 curve in Figure 3.12), has the lowest flux density and would result in the largest transformer. However, this material has the lowest coercive force and lowest core loss of any of the available materials. These factors might well be decisive in other applications.

Standard stocking factors for tap cores, wound cut cores, and laminations are listed in Table 3.3.

3.8 CORE LOSS CURVES

Choice of core material is thus based upon achieving the best characteristic for the most critical or important design parameter, with acceptable compromises on all other parameters. Figures 3.13–

Table 3.3. Standard stacking factors

Thickness, mils	SF	SF^2
0.125	0.25	0.0624
0.250	0.375	0.141
0.500	0.500	0.250
1.00	0.750	0.562
2.00	0.85	0.722
4.00	0.90	0.810
14.00	0.95	0.902
Wound Cut Core		
1.00	0.83	0.689
2.00	0.89	0.792
4.00	0.90	0.810
12.00	0.95	0.902
Lamination		
	Butt-jointed	Interleaved
4.0	0.90	0.80	0.640
6.0	0.90	0.85	0.722
14.0	0.95	0.90	0.810
18.0	0.95	0.90	0.810
25.0	0.95	0.92	0.846

SF = stacking factor

3.17 compare the core loss of some magnetic materials as a function of flux density, frequency, and material thickness.

Possibly for the first time, manufacturers' core loss data curves have been organized with the same units for all core losses. The data were digitized directly from the manufacturers data sheets. Then they were converted to metric units—tesla and watts per kilogram rather than gauss and watts per pound. These data were then put into the computer to develop a new first-order approximation in the form of the equation

$$w = k f^m B^n$$

where w is the calculated core loss density in watts per kilogram; f is the frequency in hertz; B is the flux density in tesla; and k, m, n are coefficients derived using a three-dimensional least squares fit from the digitized data. The new equations were plotted onto the same graphs as the digitized data with surprisingly close fit considering the nonlinearity of the original curves.

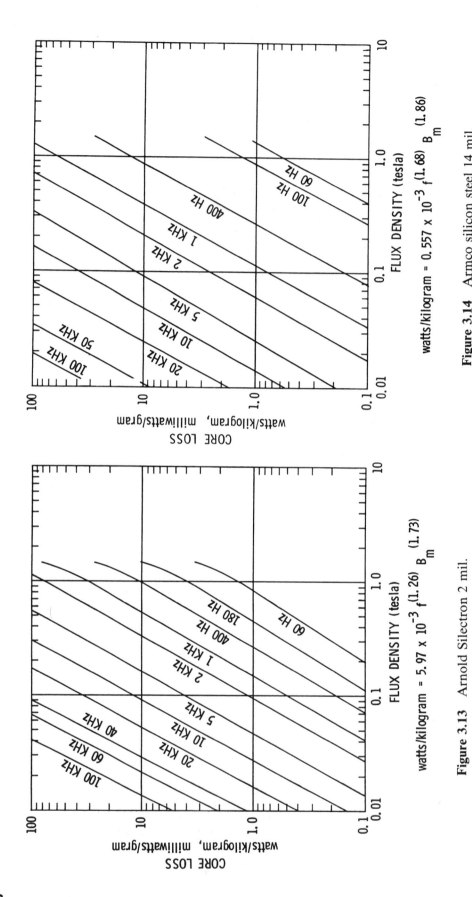

Figure 3.14 Armco silicon steel 14 mil.

$$\text{watts/kilogram} = 0.557 \times 10^{-3} f^{(1.68)} B_m^{(1.86)}$$

Figure 3.13 Arnold Silectron 2 mil.

$$\text{watts/kilogram} = 5.97 \times 10^{-3} f^{(1.26)} B_m^{(1.73)}$$

112

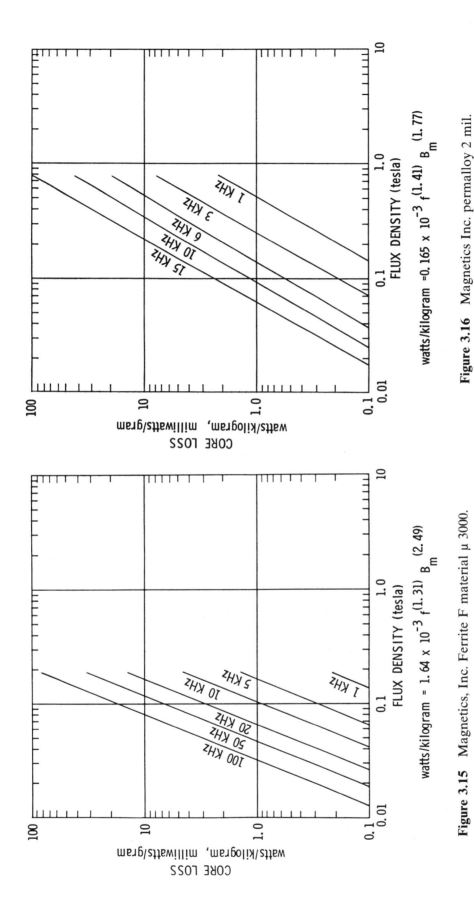

$$\text{watts/kilogram} = 0.165 \times 10^{-3} f^{(1.41)} B_m{}^{(1.77)}$$

Figure 3.16 Magnetics Inc. permalloy 2 mil.

$$\text{watts/kilogram} = 1.64 \times 10^{-3} f^{(1.31)} B_m{}^{(2.49)}$$

Figure 3.15 Magnetics, Inc. Ferrite F material μ 3000.

113

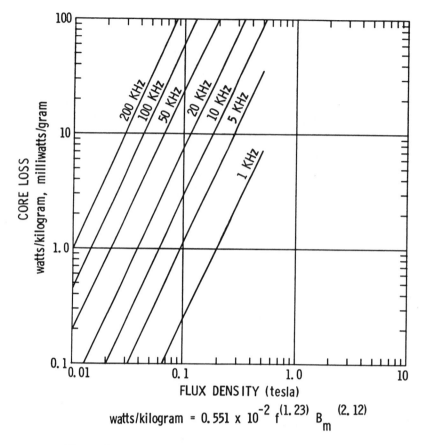

$$\text{watts/kilogram} = 0.551 \times 10^{-2} \, f^{(1.23)} \, B_m^{(2.12)}$$

Figure 3.17 Magnetics Inc. 60μ MPP powder cores.

3.9 MAGNETIC CORE TABLES

The Author has developed Tables 3.8–3.11 (pp. 116–123) to simplify and standardize the process of transformer design. They make it possible to design transformers and inductors of lighter weight and smaller volume or to optimize efficiency without going through a cut-and-try design procedure. While developed specifically for aerospace applications, the information has wider utility and can be used for the design of non-aerospace transformers as well. A great deal of information is presented for the convenience of the designer. The material is in tabular form to assist the designer in making the tradeoffs best suited for a particular application in a minimum amount of time.

Thus the circuit configuration in which the transformer is to be used must be considered by the designer when sizing the transformer.

Rather than discuss the various methods used by transformer designers, the author believes it is more useful to consider typical design problems and to work out solutions using the approach based upon the newly formulated relationships.

3.10 TRANSFORMER DESIGN USING THE AREA PRODUCT A_p APPROACH

Design work sheets using a push-pull primary and a center-tapped secondary with the following specifications.

1. Input voltage, $V_{in} = 28$ [V]
2. Output voltage, $V_o = 28$ [V]
3. Output current, $I_o = 3$ [A]
4. Frequency, $f = 20,000$ [Hz]
5. Efficiency, $\eta = 0.98$
6. Temperature rise = 25 [°C]
7. Flux density, $B_m = 0.3$ [T]
8. Core material: 80-20 nickel-iron
9. Core configuration: C core

Step 1. Calculate the transformer output power P_o allowing for a 1.0-V diode drop V_d:

$$P_o = VA \quad [W]$$

$$V = V_o + V_d \quad [V]$$

$$V = 28 + 1$$

$$P_o = (29)(3)$$

$$P_o = 87 \quad [W]$$

Step 2. Calculate the apparent power P_t:

$$P_t = P_o\left(\frac{\sqrt{2}}{\eta} + \sqrt{2}\right) \quad [W]$$

$$P_t = (87)\left(\frac{1.41}{0.98} + 1.41\right)$$

$$P_t = 249 \quad [W]$$

Table 3.4. Magnetics Inc. tape toroidal cores

Cat. no.	Bare i.d. (cm) 1	Wound o.d. (cm) 2	Wound HT (cm) 3	MPL (cm) 4	WTFE (g) 5	WTCU (g) 6
5-402	.762	1.685	.940	3.25	.630	3.15
5-107	1.079	2.144	1.099	4.24	.822	7.11
5-403	1.245	2.391	1.181	4.74	.904	10.02
5-153	.762	1.843	.940	3.49	1.342	3.35
5-056	1.079	2.310	1.099	4.49	1.727	7.54
5-057	1.397	2.783	1.257	5.48	2.107	14.01
5-000	1.079	2.639	1.099	4.99	3.838	8.37
5-063	1.397	3.104	1.257	5.98	4.600	15.40
5-002	1.461	3.198	1.289	6.18	4.753	17.13
5-011	2.324	4.535	1.721	8.97	6.899	54.42
5-076	1.372	3.451	1.397	6.48	11.244	17.57
5-061	1.702	3.581	1.727	6.98	10.737	29.10
5-106	1.702	3.907	1.562	7.48	12.979	29.01
5-007	1.372	3.451	1.562	6.48	14.952	18.95
5-004	2.324	4.535	2.038	8.97	13.799	62.08
5-168	1.702	3.581	2.045	6.98	16.106	33.20
5-167	1.384	3.780	1.568	6.98	21.475	20.61
5-094	1.397	3.448	1.892	6.48	22.428	22.38
5-029	2.324	4.852	2.038	9.47	21.852	65.91
5-318	1.702	4.239	1.727	7.98	24.551	33.20
5-181	2.007	4.382	2.197	8.47	29.316	51.99
5-032	2.324	5.176	2.038	9.97	30.674	69.75
5-188	1.689	4.247	2.038	7.98	36.827	36.84
5-030	2.946	6.124	2.362	11.96	36.796	125.16
5-026	2.324	5.176	2.356	9.97	46.011	77.41
5-038	2.324	5.176	2.699	9.97	61.348	85.69
5-035	2.946	6.124	3.023	11.96	73.592	150.78
5-425	2.946	6.774	2.692	12.96	89.714	150.28
5-055	3.556	7.083	3.340	13.96	85.899	239.00
5-169	2.921	6.815	3.023	12.96	119.618	161.75
5-017	4.813	8.993	3.981	17.95	110.450	505.61
5-031	6.071	10.912	4.610	21.93	134.940	909.88
5-012	2.946	6.124	4.293	11.96	147.185	200.05
5-103	4.775	10.280	3.975	19.94	245.390	565.58
5-128	6.045	12.168	4.635	23.93	294.493	1014.29
5-022	4.775	10.280	5.232	19.94	490.780	693.71
5-042	6.045	12.168	5.905	23.93	588.986	1221.71
5-100	5.994	12.836	7.214	24.93	1150.497	1470.48
5-112	9.741	18.555	10.471	36.89	2269.920	5291.30
5-426	9.715	20.469	10.458	39.88	3926.242	5670.91

MLT (cm) 7	Gross A_c (cm sq) 8	Gross W_a (cm sq) 9	Gross A_p (cm 4th) 10	Gross K_g (cm 5th) 11	A_t (cm sq) 12
1.94	.025	.456	.012	.000061	9.4
2.18	.025	.915	.023	.000108	14.6
2.32	.025	1.217	.030	.000131	17.9
2.06	.050	.456	.023	.000225	10.8
2.32	.050	.915	.046	.000402	16.4
2.57	.050	1.533	.077	.000606	23.2
2.57	.101	.915	.092	.001447	20.0
2.82	.101	1.533	.155	.002206	27.4
2.88	.101	1.676	.169	.002369	29.0
3.61	.101	4.243	.428	.004782	56.8
3.34	.227	1.478	.336	.009146	33.9
3.60	.202	2.275	.459	.010284	39.6
3.59	.227	2.275	.517	.013122	43.2
3.61	.302	1.478	.447	.014988	35.6
4.11	.202	4.243	.855	.016765	61.3
4.10	.302	2.275	.688	.020275	43.2
3.85	.403	1.505	.607	.025423	41.1
4.10	.454	1.533	.695	.030741	39.2
4.37	.302	4.243	1.283	.035528	68.1
4.10	.403	2.275	.917	.036045	51.2
4.62	.454	3.163	1.435	.056315	60.4
4.62	.403	4.243	1.711	.059691	75.2
4.62	.605	2.241	1.355	.070940	55.5
5.16	.403	6.819	2.750	.085926	104.4
5.13	.605	4.243	2.566	.121006	80.4
5.68	.806	4.243	3.422	.194341	86.0
6.22	.806	6.819	5.499	.285298	117.1
6.20	.907	6.819	6.187	.362265	129.4
6.77	.806	9.933	8.010	.381871	153.2
6.79	1.210	6.702	8.107	.578012	137.7
7.81	.806	18.198	14.676	.605936	239.6
8.84	.806	28.947	23.345	.851944	345.1
8.25	1.613	6.819	10.999	.860111	141.5
8.88	1.613	17.911	28.889	2.098935	294.4
9.94	1.613	28.706	46.299	3.006144	409.8
10.89	3.226	17.911	57.779	6.845035	335.0
11.97	3.226	28.706	92.599	9.983053	458.4
14.65	6.048	28.225	170.717	28.191360	549.7
19.96	8.064	74.532	601.065	97.118716	1151.4
21.51	12.903	74.144	956.696	229.570850	1330.8

Table 3.5. Magnetic metals EI and EE laminations

Cat. no.	D (cm) 1	E (cm) 2	F (cm) 3	G (cm) 4	MPL (cm) 5	HT (cm) 6	WTH (cm) 7	LT (cm) 8
1-94EI	.236	.236	.239	.396	1.7	.6	.7	1.0
1-30-31EE	.236	.236	.239	.714	2.4	1.0	.7	1.0
1-28-29EE	.317	.317	.317	.795	2.9	1.1	1.0	1.3
1-32-33EE	.356	.356	.381	.698	2.9	1.1	1.1	1.5
1-186EI	.478	.478	.478	.635	3.2	1.1	1.4	1.9
1-185EI	.478	.478	.478	.874	3.7	1.3	1.4	1.9
1-187EI	.478	.478	.478	1.113	4.1	1.6	1.4	1.9
1-188EI	.478	.478	.478	1.587	5.1	2.1	1.4	1.9
1-186-187EE	.478	.478	.478	1.748	5.4	2.2	1.4	1.9
1-186-188EE	.478	.478	.478	2.222	6.4	2.7	1.4	1.9
1-187-188EE	.478	.478	.478	2.697	7.3	3.2	1.4	1.9
1-25EIS	.635	.635	.635	.952	4.4	1.6	1.6	1.9
1-24-25EE	.635	.635	.635	1.270	5.1	1.9	1.9	2.5
1-26-38EE	.952	.952	.635	1.321	5.8	2.3	2.2	3.2
1-312EI	.795	.795	.952	1.984	7.5	2.8	2.7	3.5
1-26-27EE	.952	.952	.635	1.748	6.7	2.7	2.2	3.2
1-27-38EE	.952	.952	.635	2.113	7.4	3.1	2.2	3.2
1-375EI	.952	.952	.795	1.905	7.3	2.9	2.5	3.5
1-50EI	1.270	1.270	.635	1.905	7.6	3.2	2.5	3.8
1-21EI	1.270	1.270	.795	2.065	8.3	3.3	2.9	4.1
1-625EI	1.587	1.587	.795	2.383	9.5	4.0	3.2	4.8
1-68EI	1.748	1.748	.874	2.619	10.5	4.4	3.5	5.2
1-202EI	1.905	1.905	1.270	2.286	10.9	3.9	4.4	7.0
1-75EI	1.905	1.905	.952	2.857	11.4	4.8	3.8	5.7
1-87EI	2.222	2.222	1.113	3.335	13.3	5.6	4.4	6.7
1-100EI	2.540	2.540	1.270	3.810	15.2	6.3	5.1	7.6
1-112EI	2.857	2.857	1.430	4.288	17.2	7.1	5.7	8.6
1-125EI	3.175	3.175	1.587	4.762	19.0	7.9	6.3	9.5
1-138EI	3.492	3.492	1.748	5.240	21.0	8.7	7.0	10.5
1-150EI	3.810	3.810	1.905	5.715	22.9	9.5	7.6	11.4
1-145EI	3.683	3.683	2.349	7.620	27.3	11.4	8.4	12.1
1-36EI	4.127	4.127	3.175	6.667	27.9	10.8	10.5	14.6
1-175EI	4.445	4.445	2.222	6.680	26.7	11.1	8.9	13.3
1-19EI	4.445	4.445	4.445	7.620	33.0	12.1	13.3	17.8
1-212EI	5.397	5.397	2.700	8.098	32.4	13.5	10.8	16.2
1-225EI	5.715	5.715	2.857	8.572	34.3	14.3	11.4	17.1
1-20EI	6.350	6.350	4.762	9.525	41.3	15.9	15.9	22.2
1-3EI	7.620	7.620	3.810	11.430	45.7	19.0	15.2	22.9
1-301EI	7.620	7.620	5.715	11.430	49.5	19.0	19.0	26.7
1-4EI	10.160	10.160	5.080	15.240	61.0	25.4	20.3	30.5

Gross WTFE (kg) 9	WTCU (kg) 10	MLT (cm) 11	Gross A_c cm^2 12	Gross W_a cm^2 13	Gross A_p cm^4 14	Gross K_g cm^5 15	A_t ·cm^2 16
.001	.001	2.1	.06	.09	.01	.000056	3.0
.001	.001	2.1	.06	.17	.01	.000101	4.1
.002	.002	2.7	.10	.25	.03	.000384	6.6
.003	.003	3.0	.13	.27	.03	.000563	7.7
.006	.004	3.8	.23	.30	.07	.001652	11.0
.006	.006	3.8	.23	.42	.10	.002274	12.6
.007	.007	3.8	.23	.53	.12	.002895	14.2
.009	.010	3.8	.23	.76	.17	.004131	17.4
.009	.011	3.8	.23	.83	.19	.004547	18.5
.011	.014	3.8	.23	1.06	.24	.005783	21.7
.013	.017	3.8	.23	1.29	.29	.007019	24.8
.014	.011	4.9	.40	.60	.24	.007960	20.6
.016	.014	4.9	.40	.81	.33	.010614	23.4
.040	.019	6.2	.91	.84	.76	.044456	33.8
.036	.044	6.6	.63	1.89	1.19	.045893	47.2
.046	.025	6.2	.91	1.11	1.01	.058819	38.4
.051	.030	6.2	.91	1.34	1.22	.071129	42.4
.051	.036	6.7	.91	1.51	1.37	.074266	46.2
.094	.033	7.7	1.61	1.21	1.95	.163800	53.2
.102	.048	8.2	1.61	1.64	2.65	.298652	62.1
.183	.064	9.5	2.52	1.89	4.77	.508804	83.2
.244	.084	10.3	3.05	2.29	6.99	.825090	100.7
.302	.126	12.2	3.63	2.90	10.54	1.251557	131.7
.316	.109	11.2	3.63	2.72	9.88	1.277636	119.8
.503	.171	13.0	4.94	3.71	18.33	2.786457	163.0
.750	.254	14.8	6.45	4.84	31.22	5.458068	212.9
1.068	.360	16.5	8.17	6.13	50.06	9.890285	269.5
1.465	.492	18.3	10.08	7.56	76.21	16.795435	332.7
1.951	.654	20.1	12.20	9.16	111.69	27.152304	402.5
2.532	.853	22.0	14.52	10.89	158.04	41.638281	479.0
2.826	1.447	22.7	13.56	17.90	242.85	57.985265	600.5
3.632	2.055	27.3	17.04	21.17	360.65	90.027224	742.7
4.024	1.350	25.6	19.76	14.85	293.34	90.642741	652.0
4.978	3.922	32.6	19.76	33.87	669.22	162.443682	1066.9
7.200	2.401	30.9	29.13	21.86	636.95	240.316839	961.6
8.545	2.844	32.7	32.66	24.50	800.07	320.126244	1077.8
12.699	6.708	41.6	40.32	45.36	1829.14	709.478561	1673.4
20.255	6.763	43.7	58.06	43.55	2528.61	1344.824692	1916.1
21.943	11.534	49.7	58.06	65.32	3792.91	1774.078934	2409.7
48.013	15.918	57.8	103.23	77.42	7991.64	5706.912170	3406.4

Table 3.6. National Magnetics Corp. 2 mil C Cores

Cat. no.	D (cm) 1	E (cm) 2	F (cm) 3	G (cm) 4	MPL (cm) 5	HT (cm) 6	WTH (cm) 7	LT (cm) 8
CL-1	.635	.317	.635	1.270	4.6	1.9	1.3	1.9
CL-2	.635	.476	.635	1.587	5.6	2.5	1.3	2.2
CL-3	.952	.476	.635	1.587	5.6	2.5	1.6	2.2
CL-4	.635	.635	.635	2.222	7.4	3.5	1.3	2.5
CL-5	.952	.635	.635	2.222	7.4	3.5	1.6	2.5
CL-7	.952	.556	.794	2.540	8.1	3.7	1.7	2.7
CL-6	1.270	.635	.635	2.222	7.4	3.5	1.9	2.5
CL-124	1.270	.635	.794	2.540	8.4	3.8	2.1	2.9
CL-121	.952	.635	1.270	3.334	10.9	4.6	2.2	3.8
CL-8	.952	.952	.952	3.016	10.7	4.9	1.9	3.8
CL-9	1.270	.952	.952	3.016	10.7	4.9	2.2	3.8
CL-10	1.587	.952	.952	3.016	10.7	4.9	2.5	3.8
CL-12	1.270	1.111	1.270	2.857	11.4	5.1	2.5	4.8
CL-135	1.270	1.111	1.429	2.857	11.7	5.1	2.7	5.1
CL-11	1.905	.952	.952	3.016	10.7	4.9	2.9	3.8
CL-78	1.905	.794	.794	5.715	15.0	7.3	2.7	3.2
CL-18	1.270	1.111	1.587	3.969	14.2	6.2	2.9	5.4
CL-14	1.270	1.270	1.270	3.969	14.2	6.5	2.5	5.1
CL-15	1.587	1.270	1.270	3.969	14.2	6.5	2.9	5.1
CL-16	1.905	1.270	1.270	3.969	14.2	6.5	3.2	5.1
CL-17	2.540	1.270	1.270	3.969	14.2	6.5	3.8	5.1
CL-19	2.540	1.270	1.587	3.969	14.7	6.5	4.1	5.7
CL-79	1.905	1.270	2.222	5.080	18.0	7.6	4.1	7.0
CL-21	1.905	1.587	1.587	4.921	17.8	8.1	3.5	6.3
CL-20	2.540	1.587	1.587	3.969	15.7	7.1	4.1	6.3
CL-22	2.540	1.587	1.587	4.921	17.8	8.1	4.1	6.3
CL-24	2.540	1.587	1.905	5.874	20.1	9.0	4.4	7.0
CL-23	3.175	1.587	1.587	4.921	17.8	8.1	4.8	6.3
CL-120	2.540	1.905	2.381	6.350	22.9	10.2	4.9	8.6
CL-98	2.540	1.587	5.080	7.620	29.7	10.8	7.6	13.3
CL-53	3.492	2.222	2.381	6.826	24.9	11.3	5.9	9.2
CL-35	3.175	2.540	2.540	7.620	27.7	12.7	5.7	10.2
CL-54	5.080	1.905	1.905	10.160	29.5	14.0	7.0	7.6
CL-39	4.445	2.857	1.905	7.620	27.4	13.3	6.3	9.5
CL-29	4.445	2.857	2.540	7.620	28.7	13.3	7.0	10.8

WTFE (kg) 9	WTCU (kg) 10	MLT (cm) 11	Eff. A_c cm² 12	Gross W_a cm² 13	A_p cm⁴ 14	K_g cm⁵ 15	A_t cm² 16
.006	.009	3.3	.18	.81	.14	.003136	17.7
.011	.013	3.6	.27	1.01	.27	.008049	22.4
.017	.015	4.3	.40	1.01	.41	.015411	25.4
.020	.020	3.9	.36	1.41	.51	.018420	29.8
.030	.023	4.6	.54	1.41	.76	.035698	33.7
.029	.033	4.7	.47	2.02	.95	.038286	40.9
.040	.027	5.4	.72	1.41	1.01	.053643	37.5
.046	.041	5.7	.72	2.02	1.45	.073260	47.1
.045	.084	5.6	.54	4.23	2.28	.087935	75.7
.066	.058	5.7	.81	2.87	2.32	.131057	63.5
.088	.067	6.6	1.08	2.87	3.09	.203180	69.1
.110	.073	7.2	1.35	2.87	3.86	.289415	74.6
.110	.095	7.4	1.26	3.63	4.56	.310621	90.1
.112	.111	7.6	1.26	4.08	5.13	.338007	100.2
.131	.080	7.8	1.61	2.87	4.64	.382920	80.1
.154	.117	7.3	1.35	4.54	6.10	.452681	97.9
.136	.176	7.9	1.26	6.30	7.91	.505077	129.2
.156	.138	7.7	1.44	5.04	7.23	.540205	112.1
.195	.149	8.3	1.79	5.04	9.04	.779647	119.5
.234	.160	9.0	2.15	5.04	10.85	1.043079	126.8
.312	.183	10.2	2.87	5.04	14.46	1.624036	141.5
.323	.240	10.7	2.87	6.30	18.08	1.935620	166.4
.296	.419	10.5	2.15	11.29	24.30	2.001962	231.0
.365	.280	10.1	2.69	7.81	21.02	2.242302	183.6
.431	.254	11.4	3.59	6.30	22.60	2.855327	182.0
.487	.315	11.4	3.59	7.81	28.02	3.540605	201.9
.549	.472	11.9	3.59	11.19	40.14	4.857741	253.6
.608	.351	12.6	4.48	7.81	35.03	4.975847	220.3
.751	.712	13.2	4.31	15.12	65.09	8.465393	338.6
.814	2.318	16.8	3.59	38.70	138.85	11.829113	777.4
1.312	.912	15.8	6.91	16.25	112.24	19.648908	411.6
1.516	1.103	16.0	7.18	19.35	138.85	24.861568	467.7
1.936	1.223	17.8	8.61	19.35	166.62	32.284922	485.5
2.365	.950	18.4	11.30	14.51	164.01	40.273180	464.5
2.475	1.335	19.4	11.30	19.35	218.69	50.937499	553.2

Table 3.7. Magnetics Inc. MPP Powder Cores

Cat. no.	Bare i.d. (cm) 1	Wound o.d. (cm) 2	Wound HT (cm) 3	MPL (cm) 4	WTFE (gm) 5	WTCU (gm) 6	MLT (cm) 7
55291	.376	1.297	.696	2.18	1.800	.66	1.68
55041	.406	1.362	.711	2.38	1.900	.79	1.72
55051	.648	1.731	.926	3.12	3.100	2.44	2.08
55121	.902	2.247	1.213	4.11	6.800	6.03	2.65
55848	1.156	2.739	1.340	5.09	10.000	11.00	2.95
55059	1.288	3.055	1.533	5.67	16.000	15.53	3.35
55351	1.326	3.144	1.684	5.88	20.000	17.77	3.62
55894	1.359	3.515	1.929	6.35	36.000	21.96	4.26
55586	2.209	4.659	2.138	8.95	35.000	61.48	4.51
55071	1.879	4.367	2.151	8.15	47.000	46.19	4.68
55076	2.099	4.762	2.229	8.98	52.000	59.84	4.86
55083	2.279	5.255	2.727	9.84	92.000	84.68	5.84
55090	2.739	6.169	3.033	11.63	131.000	136.43	6.51
55716	3.039	6.724	3.005	12.73	133.000	169.06	6.55
55439	2.279	5.980	3.083	10.74	182.000	100.93	6.96
55110	3.419	7.541	3.246	14.30	176.000	233.10	7.14

Table 3.8. Magnetics Inc. Toroidal Ferrites

Cat. no.	Bare i.d. (cm) 1	Wound o.d. (cm) 2	Wount HT (cm) 3	MPL (cm) 4	WTFE (gm) 5	WTCU (gm) 6
40200-TC	.127	.318	.191	.55	.030	.02
40301-TC	.183	.443	.218	.78	.040	.04
40100-TC	.155	.456	.204	.73	.070	.03
40502-TC	.244	.514	.249	.92	.050	.07
40401-TC	.229	.600	.241	1.02	.090	.09
40503-TC	.244	.514	.376	.92	.100	.10
40601-TC	.305	.737	.304	1.30	.140	.18
40402-TC	.229	.600	.368	1.02	.170	.12
40603-TC	.305	.737	.470	1.30	.300	.25
40705-TC	.318	.932	.637	1.50	.900	.39
41003-TC	.475	1.194	.556	2.07	.820	.80
41005-TC	.475	1.194	.715	2.07	1.200	.96
41303-TC	.793	1.660	.714	3.12	1.200	2.67
40907-TC	.559	1.229	.991	2.27	.160	1.66
41506-TC	.737	1.686	.764	3.06	1.900	2.55
41206-TC	.516	1.548	.893	2.46	3.300	1.50
41306-TC	.793	1.660	1.031	3.12	2.400	3.56
41406-TC	.714	1.624	.992	2.95	2.700	2.88
41605-TC	.889	2.021	.913	3.67	3.300	4.43
42206-TC	1.372	2.886	1.321	5.42	6.900	14.60
42207-TC	1.372	2.886	1.479	5.42	8.500	15.93
42507-TC	1.550	3.304	1.568	6.17	11.600	22.09
42212-TC	1.372	2.886	1.956	5.42	13.500	19.93
42908-TC	1.900	3.834	1.699	7.32	13.800	35.41
43806-TC	1.905	4.775	1.587	8.30	26.400	41.04
43813-TC	1.905	4.775	2.160	8.30	51.700	50.29
43615-TC	2.300	4.731	2.650	9.20	44.000	77.73
44920-TC	3.180	6.471	3.178	12.30	74.600	182.35
44916-TC	3.383	6.572	3.279	12.70	75.300	206.26
44925-TC	3.180	6.471	3.495	12.30	91.000	196.65

A_c 2 (cm) 8	W_a 2 (cm) 9	A_p 4 (cm) 10	K_g 5 (cm) 11	A_t 2 (cm) 12	PERM (U) 13	MH/ 1000 TURN 14
.094	.111	.010	.000237	4.7	60.0	32.0
.100	.129	.013	.000301	5.1	60.0	32.0
.114	.330	.038	.000824	8.1	60.0	27.0
.192	.639	.123	.003553	13.6	60.0	35.0
.226	1.050	.237	.007274	18.9	60.0	32.0
.331	1.303	.431	.017039	23.8	60.0	43.0
.388	1.381	.536	.022982	26.3	60.0	51.0
.654	1.451	.949	.058299	33.8	60.0	75.0
.454	3.832	1.740	.070042	51.3	60.0	38.0
.672	2.773	1.863	.106936	48.0	60.0	61.0
.678	3.460	2.346	.130831	55.1	60.0	56.0
1.072	4.079	4.373	.321214	71.7	60.0	81.0
1.340	5.892	7.895	.649954	95.1	60.0	86.0
1.251	7.254	9.074	.692776	106.3	60.0	73.0
1.990	4.079	8.118	.928721	94.3	60.0	135.0
1.440	9.181	13.257	1.072468	130.7	60.0	75.0

MLT (cm) 7	A_c 2 (cm) 8	W_a 2 (cm) 9	A_p 4 (cm) 10	K_g 5 (cm) 11	A_t 2 (cm) 12
.41	.008	.012	.000	.000001	.3
.48	.010	.026	.000	.000002	.5
.50	.013	.019	.000	.000003	.5
.52	.011	.039	.000	.000004	.6
.59	.015	.041	.001	.000006	.8
.72.	.021	.039	.001	.000010	.9
.71	.020	.073	.001	.000016	1.2
.79	.031	.041	.001	.000020	1.1
.98	.043	.073	.003	.000055	1.6
1.37	.100	.079	.008	.000230	2.8
1.27	.073	.176	.013	.000295	3.5
1.53	.110	.176	.019	.000558	4.1
1.52	.074	.492	.036	.000707	6.3
1.90	.137	.245	.034	.000968	5.3
1.69	.112	.425	.048	.001262	6.8
2.03	.224	.208	.047	.002054	6.9
2.03	.149	.492	.073	.002150	7.9
2.03	.172	.399	.069	.002324	7.6
2.01	.158	.619	.098	.003068	9.7
2.78	.261	1.475	.385	.014437	19.6
3.04	.326	1.475	.481	.020648	21.1
3.30	.385	1.882	.725	.033805	26.4
3.80	.522	1.475	.770	.042307	25.4
3.52	.369	2.830	1.044	.043808	33.7
4.06	.581	2.840	1.650	.094358	46.6
4.98	1.110	2.840	3.152	.281103	55.2
5.28	.977	4.140	4.045	.299375	59.7
6.47	1.190	7.930	9.437	.694648	102.4
6.47	1.200	8.970	10.764	.799010	105.5
6.97	1.460	7.930	11.578	.969576	108.9

Step 3. Calculate the area product A_p:

$$A_p = \left(\frac{P_t \times 10^4}{K_f B_m f K_u K_j}\right)^{(x)} \quad [\text{cm}^4]$$

K_u, window utilization factor = 0.4

$$A_p = \left(\frac{249 \times 10^4}{(4.0)(0.3)(20,000)(0.4)(323)}\right)$$

$$A_p = 0.778 \quad [\text{cm}^4]$$

Step 4. Select a comparable area product A_p from the specified core section, and record the appropriate data:

Core CL-7

$$A_p = 0.95 \quad [\text{cm}^4]$$

$$\text{MLT} = 4.7 \quad [\text{cm}]$$

$$A_c = 0.47 \quad [\text{cm}^2]$$

$$W_a = 2.02 \quad [\text{cm}^2]$$

$$A_t = 40.9 \quad [\text{cm}^2]$$

$$\text{MPL} = 8.1 \quad [\text{cm}]$$

$$W_{tfe} = 0.029 \quad [\text{kg}]$$

Step 5. Calculate N_p, the number of primary turns each side of the center tap:

$$N_p = \frac{V_p \times 10^4}{K_f B_m f A_c} \quad \text{[turns]}$$

$$N_p = \frac{28 \times 10^4}{(4.0)(0.3)(20,000)(0.47)}$$

$$N_p = 25 \quad \text{[turns]}$$

Step 6. Calculate the primary current I_p:

$$I_p = \frac{P_o}{V_p \eta} \quad \text{[A]}$$

$$I_p = \frac{87}{(28)(0.98)}$$

$$I_p = 3.17 \quad \text{[A]}$$

Step 7. Calculate the current density J using the K_j value and (y) given for C core in Table 3.1:

$$J = K_j A_p^{(y)} \quad \text{[A/cm}^2\text{]}$$

$$J = (323)(0.95)^{-0.14}$$

$$J = 325 \quad \text{[A/cm}^2\text{]}$$

Step 8. Calculate the bare wire size $A_{w(B)}$ for the primary. In a center-tap configuration, I_p has to be multiplied by 0.707:

$$A_{w(B)} = \frac{I_p \times 0.707}{J} \quad \text{[cm}^2\text{]}$$

$$A_{w(B)} = \frac{3.17 \times 0.707}{325}$$

$$A_{w(B)} = 0.00689 \quad \text{[cm}^2\text{]}$$

Step 9. Select a wire size from the wire table (Table 6.1), column 2. Remember: If the wire area is not within 10%, take the next smallest size. Also record micro-ohms per centimeter from column 4.

AWG No. 19

$$\text{Bare, } A_{w(B)} = 0.00653 \quad [\text{cm}^2]$$

$$\mu\Omega/\text{cm} = 264$$

Step 10. Calculate the primary winding resistance. Use MLT from step 4 and micro-ohms per centimeter from step 9:

$$R_p = (\text{MLT})(N)\left(\frac{\mu\Omega}{\text{cm}}\right) \times 10^{-6} \quad [\Omega]$$

$$R_p = (4.7)(25)(264) \times 10^{-6}$$

$$R_p = 0.0310 \quad [\Omega]$$

Step 11. Calculate the primary copper loss P_p:

$$P_p = I_p^2 R_p \quad [\text{W}]$$

$$P_p = (3.17)^2(0.031)$$

$$P_p = 0.311 \quad [\text{W}]$$

Step 12. Calculate N, the number of secondary turns each side of the center tap:

$$N_s = \frac{N_p V_s}{V_p} \quad [\text{turns}]$$

$$N_s = \frac{(25)(29)}{28}$$

$$N_s = 26 \quad [\text{turns}]$$

Step 13. Calculate the bare wire size $A_{w(B)}$ for the secondary. In a center-tap configuration, I_o has to be multiplied by 0.707:

$$A_{w(B)} = \frac{I_o(0.707)}{J} \quad [\text{cm}^2]$$

$$A_{w(B)} = \frac{(3)(0.707)}{325}$$

$$A_{w(B)} = 0.00653 \quad [\text{cm}^2]$$

Step 14. Select a wire size from the wire table (Table 6.1), column 2. If the wire area is not within 10%, take the next smallest size. Also record micro-ohms per centimeter from column 4.

AWG No. 19

$$\text{Bare,} \, A_{w(B)} = 0.00653 \quad [\text{cm}^2]$$

$$\mu\Omega/\text{cm} = 264$$

Step No. 15. Calculate the secondary winding resistance. Use MLT from step 4 and micro-ohms per centimeter from step 14:

$$R_s = (\text{MLT})(N)\left(\frac{\mu\Omega}{\text{cm}}\right) \times 10^{-6} \quad [\Omega]$$

$$R_s = (4.7)(26)(264) \times 10^{-6}$$

$$R_s = 0.0322 \quad [\Omega]$$

Step 16. Calculate the secondary copper loss P_s:

$$P_s = I_o^2 R_s \quad [\text{W}]$$

$$P_s = (3.0)^2 (0.0322)$$

$$P_s = 0.290 \quad [\text{W}]$$

Step 17. Calculate the transformer total copper loss:

$$P_{cu} = P_p + P_s \quad [\text{W}]$$

$$P_{cu} = 0.311 + 0.290$$

$$P_{cu} = 0.601 \quad [\text{W}]$$

Step 18. Calculate what the combined losses P_Σ have to be to meet the efficiency specification:

$$P_\Sigma = \frac{P_o}{\eta} - P_o \quad [\text{W}]$$

$$P_\Sigma = \frac{87}{0.98} - 87$$

$$P_\Sigma = 1.77 \quad [\text{W}]$$

Step 19. Calculate the iron losses P_{fe}:

$$P_{fe} = P_\Sigma - P_{cu} \quad [\text{W}]$$

$$P_{fe} = 1.77 - 0.601$$

$$P_{fe} = 1.17 \quad [\text{W}]$$

Step 20.   Calculate the core loss P_{fe} in milliwatts per gram:

$$\text{Core loss} = \frac{P_{fe}}{W_{tfe} \times 10^{-3}} \quad [\text{mW/g}]$$

$$\text{Core loss} = \frac{1.17}{29 \times 10^{-3}}$$

$$\text{Core loss} = 40 \quad [\text{mW/g}]$$

Step 21.   Using the appropriate core loss curves find the magnetic material that comes the closest to meeting the requirement. Then calculate the watts per kilogram. Core weight is found in step 4:

$$\text{W/kg} = K f^{(m)} B_m^{(n)}$$

$$\text{W/kg} = (0.165)(20{,}000)^{1.41}(0.3)^{1.77} \times 10^{-3}$$

$$\text{W/kg} = 22.7$$

$$P_{fe} = (\text{W/kg}) W_{tfe} \quad [\text{W}]$$

$$P_{fe} = (22.7)(0.029)$$

$$P_{fe} = 0.659 \quad [\text{W}]$$

Step 22.   Calculate ψ, the watts per unit area. The surface area A_t is found in step 4:

$$\psi = \frac{P_{cu} + P_{fe}}{A_t} \quad [\text{W/cm}^2]$$

$$\psi = \frac{1.27}{40.9}$$

$$\psi = 0.0309 \quad [\text{W/cm}^2]$$

3.11 TRANSFORMER DESIGN USING THE CORE GEOMETRY K_g APPROACH

Design work sheets for a push-pull primary and a center-tapped secondary with the following specifications:

1. Input voltage, V_{in} = 22 [V]
2. Output voltage, V_o = 10 [V]
3. Output current, I_o = 10 [A]
4. Frequency, f = 10,000 [Hz]
5. Efficiency, η = 0.98
6. Regulation, α = 1.0 [%]
7. Flux density, B_m = 0.3 [T]
8. Core material: 80-20 nickel-iron
9. Core configuration: 2-mil C core

Step 1. Calculate the transformer output power P_o allowing for a 1.0-V diode drop V_d:

$$P_o = VA \quad [W]$$

$$V = V_o + V_d \quad [V]$$

$$V = 10 + 1$$

$$P_o = (11)(10)$$

$$P_o = 110 \quad [W]$$

Step 2. Calculate the apparent power P_t:

$$P_t = P_o\left(\frac{\sqrt{2}}{\eta} + \sqrt{2}\right) \quad [W]$$

$$P_t = (110)\left(\frac{1.41}{0.98} + 1.41\right)$$

$$P_t = 313 \quad [W]$$

Step 3.   Calculate the electrical conditions:

$$K_e = 0.145 K_f^2 f^2 B_m^2 \times 10^{-4}$$

$$K_f = 4.0, \text{ square wave}$$

$$K_e = (0.145)(4.0)^2(10,000)^2(0.3)^2 \times 10^{-4}$$

$$K_e = 2088$$

Step 4.   Calculate the core geometry coefficient K_g:

$$K_g = \frac{P_t}{2K_e\alpha} \quad [\text{cm}^5]$$

$$K_g = \frac{313}{(2)(2088)(1)}$$

$$K_g = 0.0749 \quad [\text{cm}^5]$$

Step 5.   Select a comparable core geometry K_g from the specified core section, and record the appropriate data:

Core CL-121

$$K_g = 0.0879 \quad [\text{cm}^5]$$

$$A_p = 2.28 \quad [\text{cm}^4]$$

$$MLT = 5.6 \quad [\text{cm}]$$

$$A_c = 0.54 \quad [\text{cm}^2]$$

$$W_a = 4.23 \quad [\text{cm}^2]$$

$$A_t = 75.7 \quad [cm^2]$$

$$MPL = 10.9 \quad [cm]$$

$$W_{tfe} = 0.045 \quad [kg]$$

Step 6. Calculate N_p, the number of primary turns each side of the center tap:

$$N_p = \frac{V_p \times 10^4}{K_f B_m f A_c} \quad [turns]$$

$$N_p = \frac{22 \times 10^4}{(4.0)(0.3)(10,000)(0.54)}$$

$$N_p = 34 \quad [turns]$$

Step 7. Calculate the primary current I_p:

$$I_p = \frac{P_o}{V_p \eta} \quad [A]$$

$$I_p = \frac{110}{(22)(0.98)}$$

$$I_p = 5.10 \quad [A]$$

Step 8. Calculate the current density J:

$$J = \frac{P_t \times 10^4}{K_f K_u f B_m A_p} \quad [A/cm^2]$$

Window utilization factor $K_u = 0.4$

$$J = \frac{313 \times 10^4}{(4.0)(0.4)(10,000)(0.3)(2.28)}$$

$$J = 286 \quad [\text{A/cm}^2]$$

Step 9. Calculate the bare wire size $A_{w(B)}$ for the primary. In a center-tap configuration, I_p has to be multiplied by 0.707:

$$A_{w(B)} = \frac{(I_p)(0.707)}{J} \quad [\text{cm}^2]$$

$$A_{w(B)} = \frac{(5.10)(0.707)}{286}$$

$$A_{w(B)} = 0.126 \quad [\text{cm}^2]$$

Step 10. Select a wire size from Table 6.1, column 2. Remember: If the wire area is not within 10%, take the next smallest size. Also record micro-ohms per centimeter from column 4.

AWG No. 16

$$\text{Bare}, A_{w(B)} = 0.013 \quad [\text{cm}^2]$$

$$\mu\Omega/\text{cm} = 132$$

Step 11. Calculate the primary winding resistance. Use MLT from step 5 and micro-ohms per centimeter from step 10:

$$R_p = \text{MLT}(N)\left(\frac{\mu\Omega}{\text{cm}}\right) \times 10^{-6} \quad [\Omega]$$

$$R_p = (5.6)(34)(132) \times 10^{-6}$$

$$R_p = 0.0251 \quad [\Omega]$$

Step 12. Calculate the primary copper loss P_p:

$$P_p = I_p^2 R_p \quad [\text{W}]$$

$$P_p = (5.10)^2(0.0251)$$

$$P_p = 0.653 \quad [\text{W}]$$

Step 13. Calculate N_s, the number of secondary turns each side of the center tap:

$$N_s = \frac{N_p V_s}{V_p}\left(1 + \frac{\alpha}{100}\right) \quad [\text{turns}]$$

$$N_s = \frac{(34)(11)}{22}\left(1 + \frac{1}{100}\right)$$

$$N_s = 17 \quad [\text{turns}]$$

Step 14. Calculate the bare wire size $A_{w(B)}$ for the secondary. In a center-tap configuration, I_o has to be multiplied by 0.707:

$$A_{w(B)} = \frac{I_o(0.707)}{J} \quad [\text{cm}^2]$$

$$A_{w(B)} = \frac{(10)(0.707)}{286}$$

$$A_{w(B)} = 0.0247 \quad [\text{cm}^2]$$

Step No. 15. Select a wire size from Table 6.1, column 2. If the wire area is not within 10%, take the next smallest size. Also record micro-ohms per centimeter from column 4.

AWG No. 13 (use two No. 16 or four No. 19)

$$\text{Bare}, A_{w(B)} = 0.0262 \quad [\text{cm}^2]$$

$$\mu\Omega/\text{cm} = 65.6$$

Step 16. Calculate the secondary winding resistance. Use MLT from step 5 and micro-ohms per centimeter from step 15:

$$R_s = (\text{MLT})(N)\left(\frac{\mu\Omega}{\text{cm}}\right) \times 10^{-6} \quad [\Omega]$$

$$R_s = (5.6)(17)(66) \times 10^{-6}$$

$$R_s = 0.00628 \quad [\Omega]$$

Step 17. Calculate the secondary copper loss P_s:

$$P_s = I_o^2 R_s \quad [\text{W}]$$

$$P_s = (10)^2(0.0063)$$

$$P_s = 0.628 \quad [\text{W}]$$

Step 18. Calculate the transformer regulation, α:

$$\alpha = \frac{P_{\text{cu}} \times 100}{P_o} \quad [\%]$$

$$P_{\text{cu}} = P_p + P_s \quad [\text{W}]$$

$$P_{\text{cu}} = 0.653 + 0.628$$

$$P_{\text{cu}} = 128 \quad [\text{W}]$$

$$\alpha = \frac{1.28 \times 100}{110}$$

$$\alpha = 1.16 \quad [\%]$$

Step 19. Calculate what the combined losses P_Σ have to be to meet the efficiency specifications:

$$P_\Sigma = \frac{P_o}{\eta} - P_o \quad [W]$$

$$P_\Sigma = \frac{110}{0.98} - 110$$

$$P_\Sigma = 2.24 \quad [W]$$

Step 20. Calculate the iron losses P_{fe}:

$$P_{fe} = P_\Sigma - P_{cu} \quad [W]$$

$$P_{fe} = 2.24 - 1.28$$

$$P_{fe} = 0.96 \quad [W]$$

Step 21. Calculate the core loss P_{fe} in watts per kilogram:

$$kg = \frac{P_{fe}}{W_{tfe}}$$

$$kg = \frac{0.96}{0.045}$$

$$kg = 21.3$$

Step 22. Using the appropriate core loss curves, find the magnetic material that comes the closest to meeting the requirement. Then calculate the watts per kilogram. Core weight is found in step 4:

$$W/kg = K f^{(m)} B_m^{(n)}$$

$$W/kg = (0.165)(10,000)^{1.41}(0.3)^{1.77} \times 10^{-3}$$

$$W/kg = 8.55$$

$$P_{fe} = (W/kg)(W_{tfe}) \quad [W]$$

$$P_{fe} = (8.55)(0.045)$$

$$P_{fe} = 0.385 \quad [W]$$

REFERENCES

1. McLyman, C., *Design Parameters of Toroidal and Bobbin Magnetics,* Technical Memorandum 33-651, pages 12–15, Jet Propulsion Laboratory, Pasadena, Calif.
2. Blume, L.F., *Transformer Engineering,* John Wiley & Sons Inc., New York, N.Y. 1938. Pages 272–282.
3. Terman, F.E., *Radio Engineers Handbook,* McGraw-Hill Book Co., Inc., New York 1943. Pages 28–37.
4. Reed, E.G., *The Essentials of Transformers Practice,* D. Van Nostrand Co. Inc., New York 1927. Pages 77–90.

APPENDIX A. TRANSFORMER POWER-HANDLING CAPABILITY

The relationship between the power-handling capability of a transformer and the area product A_p can be derived as follows.

A form of the Faraday law of electromagnetic induction much used by transformer designers states

$$E = K B_m A_c N f \times 10^{-4} \qquad (3.A-1)$$

(The constant K is taken as 4 for square wave and as 4.44 for sine wave operation.)

It is convenient to restate this expression as

$$NA_c = \frac{E \times 10^4}{4B_m f} \tag{3.A-2}$$

for the following manipulation.

By definition, the window utilization factor is

$$K_u = \frac{NA_w}{W_a} \tag{3.A-3}$$

and this may be restated as

$$N = \frac{K_u W_a}{A_w} \tag{3.A-4}$$

If both sides of the equation are multiplied by A_c, then

$$NA_c = \frac{K_u W_a A_c}{A_w} \tag{3.A-5}$$

From Equation (3.A-2):

$$\frac{K_u W_a A_c}{A_w} = \frac{E \times 10^4}{4B_m f} \tag{3.A-6}$$

Solving for $W_a A_c$ we get

$$W_a A_c = \frac{E A_w \times 10^4}{4B_m f K_u} \tag{3.A-7}$$

By definition, current density J is given in amperes per square centimeter (A/cm^2), expressed as

$$J = \frac{I}{A_w} \tag{3.A-8}$$

Rearranging, this may be restated as

$$A_w = \frac{I}{J} \tag{3.A-9}$$

It will be remembered that transformer efficiency is defined as

$$\eta = \frac{P_o}{P_{\text{in}}} \quad \text{and} \quad P_{\text{in}} = EI \tag{3.A-10}$$

Rewriting Equation (3.A-7) as

$$EA_w = 4B_m f K_u W_a A_c \times 10^{-4} = \frac{EI}{J} \tag{3.A-11}$$

and since

$$\frac{EI}{J} = \frac{P_{\text{in}}}{J} = \frac{P_o}{J\eta} \tag{3.A-12}$$

then

$$W_a A_c \bigg|_{\text{total}} = W_a A_c \bigg|_{\text{primary}} + W_a A_c \bigg|_{\text{secondary}}$$

$$W_a A_c \bigg|_{\text{total}} = \frac{P_o \times 10^4}{\eta J 4 B_m f K_u} + \frac{P_o \times 10^4}{4 B_m f K_u J} = \frac{P_o \times 10^4}{4 B_m f K_u J}\left(\frac{1}{\eta} + 1\right) \tag{3.A-13}$$

and since

$$P_t = \frac{P_o}{\eta} + P_o \tag{3.A-14}$$

then

$$W_a A_c = \frac{P_t \times 10^4}{4 B_m f K_u J} \tag{3.A-15}$$

$$A_p = \frac{P_t \times 10^4}{4 B_m f J K_u} \tag{3.A-16}$$

Combining this with the equation from Table 3.1,

$$J = K_j A_p^{-0.14} \tag{3.A-17}$$

yields

$$A_p = \frac{P_t \times 10^4}{4 B_m f K_u (K_j A_p^{-0.14})} \tag{3.A-18}$$

$$A_p^{0.86} = \frac{P_t \times 10^4}{4 B_m f K_u K_j} \tag{3.A-19}$$

$$A_p = \left(\frac{P_t \times 10^4}{4 B_m f K_u K_j} \right)^{1.16} \quad [\text{cm}^4] \tag{3.A-20}$$

APPENDIX B. TRANSFORMERS DESIGNED FOR A GIVEN REGULATION

Although most transformers are designed for a given temperature rise, they can also be designed for a given regulation. The regulation and power-handling ability of a core is related to two constants, K_g and K_e, by the equation

$$P_t = 2 K_g K_e \alpha \tag{3.B-1}$$

where

$$\alpha = \text{regulation, \%.}$$

The constant K_g is determined by the core geometry:

$$K_g = f(A_c, W_a, \text{MLT}) \tag{3.B-2}$$

The constant K_e is determined by the magnetic and electric operating conditions:

$$K_e = f(f, B_m) \tag{3.B-3}$$

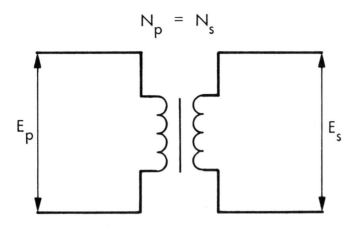

Figure 3.B-1. Isolation transformer.

The derivation of the specific functions for K_g and K_e is as follows: First assume two-winding transformers with equal primary and secondary regulation, schematically shown in Figure 3.B-1. The primary winding has a resistance R_p ohms, and the secondary winding has a resistance R_s ohms:

$$\alpha = \frac{\Delta E_p}{E_p} \times 100 + \frac{\Delta E_s}{E_s} \times 100 \qquad (3.B\text{-}4)$$

$$\Delta E_p = R_p I_p \qquad (3.B\text{-}5)$$

$$\Delta E_s = R_s I_s \qquad (3.B\text{-}6)$$

$$\alpha = 2\,\frac{R_p I_p}{E_p} \times 100 \qquad (3.B\text{-}7)$$

Multiply the numerator and denominator by E_p:

$$\alpha = 200\,\frac{R_p I_p}{E_p}\left(\frac{E_p}{E_p}\right) \qquad (3.B\text{-}8)$$

$$\alpha = 200\,\frac{R_p VA}{E_p^2} \qquad (3.B\text{-}9)$$

From the resistivity formula, it is easily shown that

$$\text{Primary:} \quad R_p = \frac{\text{MLT}\,N_p^2}{W_a K_a}\,\rho \tag{3.B-10}$$

where $\rho = 1.724 \times 10^{-6}$ ohm cm
K_a is the window utilization factor (primary)
K_b is the window utilization factor (secondary)

and

$$K_a = K_b$$

Faraday's law expressed in metric units is

$$E_p = K f N A_c B_m \times 10^{-4} \tag{3.B-11}$$

where $K = 4.0$ for a square wave and 4.44 for a sine wave.

Substituting Equations (3.B-10) and (3.B-11) for R_p and E_p in Equation 3.B-12,

$$VA = \frac{E_p^2}{200 R_p}\,\alpha \tag{3.B-12}$$

$$\text{Primary:} \quad VA = \frac{(K f N_p A_c B_m \times 10^{-4})(K f N_p A_c B_m \times 10^{-4})}{200 \times (\text{MLT}\,N_p^2 / W_a K_a)^\rho}\,\alpha \tag{3.B-13}$$

$$\text{Primary:} \quad VA = \frac{K^2 f^2 A_c^2 B_m^2 W_a K_a \times 10^{-10}}{2\,\text{MLT}\,\rho}\,\alpha \tag{3.B-14}$$

Inserting 1.724×10^{-6} for ρ,

$$\text{Primary:} \quad VA = \frac{0.29 K^2 f^2 A_c^2 B_m^2 W_a K_a \times 10^{-4}}{\text{MLT}}\,\alpha \tag{3.B-15}$$

Let

$$\text{Primary:} \quad K_e = 0.29K^2f^2B_m^2 \times 10^{-4} \quad \text{(3.B-16)}$$

and

$$\text{Primary:} \quad K_g = \frac{W_a K_a A_c^2}{\text{MLT}} \quad [\text{cm}^5] \quad \text{(3.B-17)}$$

The total transformer window utilization factor is then

$$K_a + K_b = K_u \quad \text{(3.B-18)}$$

$$K_a = \frac{K_u}{2}$$

When this value for K_a is put into Equation 3.B-15, then

$$VA = K_e K_g \alpha$$

where

$$K_e = 0.145K^2f^2B_m^2 \times 10^{-4} \quad \text{(3.B-19)}$$

The above VA is the primary power, and the window utilization factor K_u includes both the primary and secondary coils.

$$K_g = \frac{W_a K_u A_c^2}{\text{MLT}} \quad [\text{cm}^5] \quad \text{(3.B-20)}$$

Regulation of a transformer is related to the copper loss as shown in Equation (3.21):

$$\alpha = \frac{P_{cu}}{P_o} \times 100 \quad [\%] \quad \text{(3.B-21)}$$

The total VA of the transformer is

$$\text{Primary:} \qquad VA = K_e K_g \alpha$$

plus

$$\text{Secondary:} \qquad VA = K_e K_g \alpha$$

$$P_t = (\text{primary}) K_e K_g \alpha + (\text{secondary}) K_e K_g \alpha$$

$$P_t = 2 K_e K_g \alpha \qquad\qquad\qquad\qquad\qquad\qquad (3.B\text{-}22)$$

Chapter 4

Simplified Cut Core Inductor Design

4.1 INTRODUCTION

Designers have used various approaches in arriving at suitable inductor designs. For example, in many cases a rule of thumb used for dealing with current density is that a good working level is 1000 circular mils per ampere. This is satisfactory in many instances; however, the wire size used to meet this requirement may produce a heavier and bulkier inductor than desired or required. The information presented herein will make it possible to avoid the use of this and other rules of thumb and to develop a more economical and a better design.

4.2 CORE MATERIAL

Designers have routinely tended to specify Molypermalloy powder core materials for filter inductors used in high-frequency power converters and pulse-width-modulated (PWM) switched regulators because of the availability of manufacturers' literature containing tables, graphs, and examples that simplify the design task. Use of these cores may not result in an inductor design optimized for size and weight. For example, as shown in Figure 4.1, Molypermalloy powder cores operating with a dc bias of 0.3 T have only about 80% of the original inductance, with very rapid falloff at higher densities. In contrast, the steel core has approximately four times the useful flux density capability while retaining 90% of the original inductance at 1.2 T.

There are significant advantages to be gained by the use of C cores and cut toroids fabricated from grain-oriented silicon steel, despite such disadvantages as the need for banding and gapping materials, banding tools, mounting brackets, and winding mandrels.

Grain-oriented silicon steels provide greater flexibility in the design of high-frequency inductors because the air gap can be adjusted to any desired length and because the relative permeability

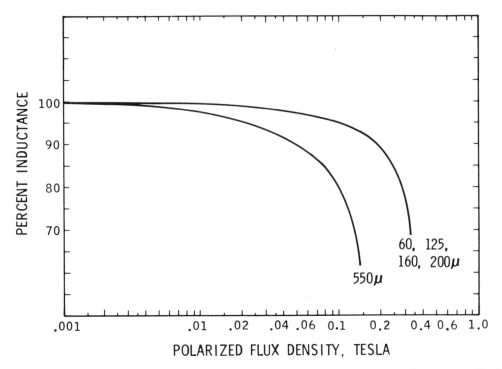

Figure 4.1. Inductance vs. dc bias for Molypermalloy cores. (*Courtesy of Magnetics Inc.*)

is high even at high dc flux density. Such steels can develop flux densities of 1.6 T, with useful linearity to 1.2 T. Molypermalloy cores [Ref. 1] carrying dc current, on the other hand, have useful flux density capabilities to only about 0.3 T.

4.3 RELATIONSHIP OF K_g TO INDUCTOR ENERGY-HANDLING CAPABILITY

Inductors, like transformers, are designed for a given temperature rise. They can also be designed for a given regulation. The regulation and energy-handling ability of a core are related to the two constants K_g and K_e by the equation:*

$$\text{Regulation } \alpha = \frac{(\text{Energy})^2}{K_g K_e} \quad [\%] \tag{4.1}$$

The constant K_g is determined by the core geometry:

$$K_g = \frac{W_a A_c^2 K_u}{\text{MLT}} \quad [\text{cm}^5] \tag{4.2}$$

---

*The derivations of Equations (4.1)–(4.3) are set forth in detail in Appendix B at the end of this chapter.

The constant K_e is determined by the magnetic and electrical operating conditions:

$$K_e = 0.145 P_o B_{max}^2 \times 10^{-4} \qquad (4.3)$$

where P_o is output power, and

$$B_{max} = B_{dc} + \frac{B_{ac}}{2}$$

From these equations it can be seen that flux density is the predominant factor governing size.

4.4 RELATIONSHIP OF A_p TO INDUCTOR ENERGY-HANDLING CAPABILITY

The energy-handling capability of a core is related to its area product A_p by an equation that may be stated as follows:*

$$A_p = \left(\frac{2(\text{Energy}) \times 10^4}{B_m K_u K_j} \right)^{1.14} \quad [\text{cm}^4] \qquad (4.4)$$

where K_j is the current density coefficient (see Chapter 2)
K_u is the window utilization factor (see Chapter 6)
B_m is the flux density, tesla
Energy is given in watt-seconds

From this it can be seen that factors such as flux density, the window utilization factor K_u (which defines the maximum space that may be occupied by the copper in the window), and the constant K_j (which is related to temperature rise) all have an influence on the inductor area product. The constant K_j is a parameter that gives the designer control of the copper loss. Derivation is set forth in detail in Chapter 2.

4.5 FUNDAMENTAL CONSIDERATIONS

The design of a linear reactor depends upon four related factors:

1. Desired inductance
2. Direct current
3. Alternating current ΔI
4. Power loss and temperature rise

---

*The derivation of Equation (4.4) is set forth in detail in Appendix A at the end of this chapter.

Table 4.1. Magnetic saturation values for selected materials

Material		Flux density, T
Magnesil	3% Si, 97% Fe	1.6
Orthonol	50% Ni, 50% Fe	1.5
48 Alloy	48% Ni, 50% Fe	1.2
Permalloy	79% Ni, 17% Fe, 4% Mo	0.75

With these requirements established, the designer must determine the maximum values for B_{dc} and B_{ac} that will not produce magnetic saturation, and must make tradeoffs that will yield the highest inductance for a given volume. The choice of core material dictates the maximum flux density that can be tolerated for a given design. Magnetic saturation values for some core materials are given in Table 4.1.

It should be remembered that maximum flux density depends upon $B_{dc} + B_{ac}$ in the manner shown in Figure 4.2, which can be expressed as

$$B_{max} = B_{dc} + B_{ac} \quad [T]$$

$$B_{dc} = \frac{0.4\pi N I_{dc} \times 10^{-4}}{l_g + l_m/\mu_r} \quad [T] \tag{4.5}$$

$$B_{ac} = \frac{0.4\pi N(\Delta I/2) \times 10^{-4}}{l_g + l_m/\mu_r} \quad [T] \tag{4.6}$$

Combining Eqs. (4.5) and (4.6),

$$B_{max} = \frac{0.4\pi N I_{dc} \times 10^{-4}}{l_g + l_m/\mu_r} + \frac{0.4\pi N(\Delta I/2) \times 10^{-4}}{l_g + l_m/\mu_r} \quad [T] \tag{4.7}$$

The inductance of an iron-core inductor carrying direct current and having an air gap may be expressed as

$$L = \frac{0.4\pi N^2 A_c \times 10^{-8}}{l_g + l_m/\mu_r} \quad [henry] \tag{4.8}$$

This shows that inductance is dependent on the effective length of the magnetic path, which is the sum of the air gap length l_g and the ratio of the core mean length to relative permeability, l_m/μ_r.

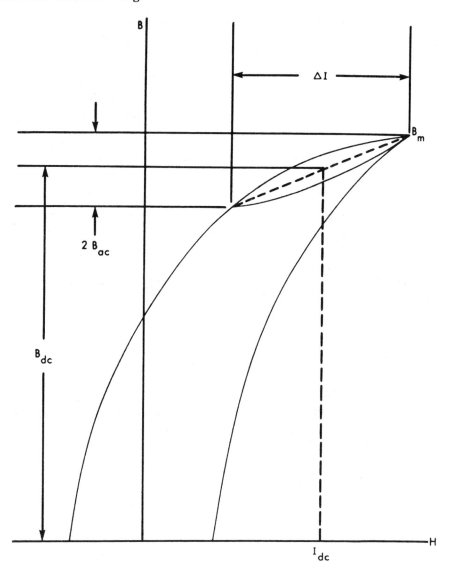

Figure 4.2. Flux density vs. $I_{dc} + \Delta I$.

When the core air gap l_g is large compared to the ratio l_m/μ_r because of high relative permeability μ_r, variations in μ_r do not substantially affect the total effective magnetic path length or the inductance. The inductance equation then reduces to

$$L = \frac{0.4\pi N^2 A_c \times 10^{-8}}{l_g} \quad \text{[henry]} \tag{4.9}$$

Final determination of the air gap size requires consideration of the effect of fringing flux, which is a function of gap dimension, the shape of the pole faces, and the shape, size, and location of the winding. Its net effect is to shorten the air gap.

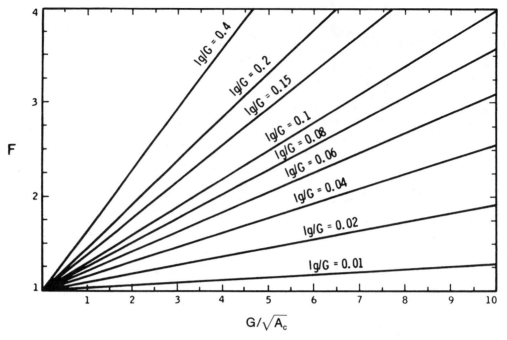

Figure 4.3. Increase of reactor inductance with flux fringing at the gap.

Fringing flux decreases the total reluctance of the magnetic path and therefore increases the inductance by a factor F to a value greater than that calculated from Equation (4.9). Fringing flux [Ref. 2] is a larger percentage of the total for larger gaps. The fringing flux factor is

$$F = 1 + \frac{l_g}{\sqrt{A_c}} \ln\left(\frac{2G}{l_g}\right) \qquad (4.10)$$

where G is a dimension defined in Chapter 2. (This equation is also valid for laminations.)

Equation (4.10) is plotted in Figure 4.3.

The inductance L computed in Equation (4.9) does not include the effect of fringing flux. The value of inductance L' corrected for fringing flux is

$$L' = \frac{0.4\pi N^2 A_c F \times 10^{-8}}{l_g} \quad \text{[henry]} \qquad (4.11)$$

Effective permeability may be calculated from the following expression:

$$\mu_\Delta = \frac{\mu_m}{1 + (l_g/l_m)\mu_m} \qquad (4.12)$$

where μ_m is the core material permeability.

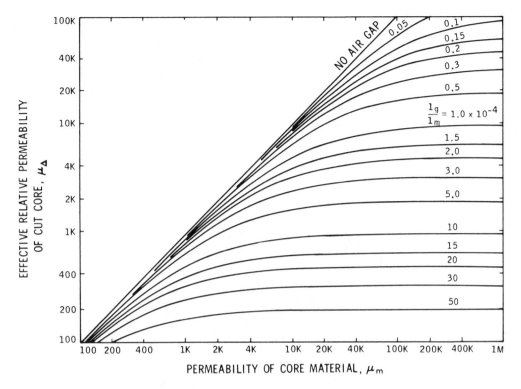

Figure 4.4. Effective permeability of cut core vs. permeability of the material. (*Courtesy of Arnold Engineering Co.*)

Figure 4.4 shows plotted curves of effective design permeability for a butt core joint structure versus material permeabilities ranging from 100 to 1 million for values of l_g/l_m of 0–0.005. The variation in effective permeability as a function of core geometry is shown in the curves plotted in Figure 4.5.

After establishing the required inductance and the dc bias current that will be encountered, dimensions can be determined. This requires consideration of the energy-handling capability, which is controlled by the area product A_p. The energy-handling capability of a core is derived from

$$\frac{LI^2}{2} = \text{Energy} \quad \text{[watt-seconds]} \tag{4.13}$$

and

$$A_p = \left(\frac{2(\text{Energy}) \times 10^4}{B_m K_u K_j} \right)^{1.16} \quad \text{[cm}^4\text{]} \tag{4.14}$$

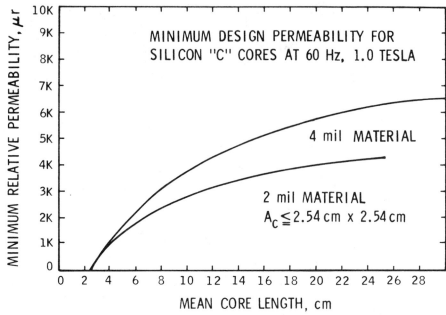

Figure 4.5. Minimum design permeability for silicon C cores at 60 Hz and 1.0 T. (*Courtesy of Arnold Engineering Co.*)

where B_m is the maximum flux density $B_{dc} + B_{ac}$

  $K_u = 0.4$ (Chapter 6)
  $K_j =$ (See Chapter 2)
Energy is given in watt-seconds

Rather than discuss the various methods used by transformer designers, the author believes it is more useful to consider typical design problems and to work out solutions using the approach based upon the newly formulated relationships. Sections 4.6 and 4.7 present two worked-out examples of gapped inductor designs.

Inductors designed in this handbook are banded together with phosphor bronze banding material or held together with aluminum brackets. The use of steel banding material or brackets that bridge the gap is called *shorting the gap*. When the gap is shorted, the inductance will increase from the calculated valve.

Much of the information that the designer needs can only be found in a scattered variety of texts and other literature. To make this information more conveniently available, helpful data has been gathered together and reproduced in Appendix C at the end of this chapter which contains 20 tables and 22 figures. Specific information can be readily located by refering to the index at the end of the book.

4.6 GAPPED INDUCTOR DESIGN USING THE AREA PRODUCT A_p APPROACH: AN EXAMPLE

Design work sheets for a gapped inductor with the following specifications:

1. Inductance, $L = 0.002$ [H]
2. dc current, $I_o = 5$ [A]
3. ac current, $\Delta I = 0.5$ [A]
4. Ripple frequency, $f = 20,000$ [Hz]
5. Flux density, $B_m = 1.4$ [T]
6. Temperature rise = 25 [°C]
7. Core material: silicon
8. Core configuration: C core

Step 1. Calculate the energy-handling capability:

$$\text{Energy} = \frac{LI^2}{2} \quad [\text{W-s}]$$

$$I = I_o + \frac{\Delta I}{2} \quad [\text{A}]$$

$$I = 5 + \frac{0.5}{2}$$

$$I = 5.25 \quad [\text{A}]$$

$$\text{Energy} = \frac{(0.002)(5.25)^2}{2}$$

$$\text{Energy} = 0.0276 \quad [\text{W-s}]$$

Step 2. Calculate the area product A_p:

$$A_p = \left(\frac{2(\text{Energy}) \times 10^4}{B_m K_u K_j} \right)^{(x)} \quad [\text{cm}^4]$$

Window utilization factor $K_u = 0.4$

$$A_p = \left(\frac{2(0.0276) \times 10^4}{(1.4)(0.4)(395)} \right)^{1.16}$$

$$A_p = 2.88 \quad [\text{cm}^4]$$

Step 3. Select a comparable area product A_p from the specified core section in Chapter 3, and record the appropriate data.

Core CL-9

$$A_p = 3.09 \quad [\text{cm}^4]$$

$$\text{MLT} = 6.6 \quad [\text{cm}]$$

$$A_c = 1.08 \quad [\text{cm}^2]$$

$$W_a = 2.87 \quad [\text{cm}^2]$$

$$A_t = 69.1 \quad [\text{cm}^2]$$

$$\text{MPL} = 10.7 \quad [\text{cm}]$$

$$W_{tfe} = 88 \quad [\text{g}]$$

$$G = 3.016 \quad [\text{cm}]$$

Step 4. Calculate the current density J using K_j from Table   . :

$$J = K_j A_p^{(y)} \quad [\text{A/cm}^2]$$

$$J = (369)(3.09)^{-0.14}$$

$$J = 315 \quad [\text{A/cm}^2]$$

Step 5. Calculate the bare wire size $A_{w(B)}$:

$$A_{w(B)} = \frac{I_o + \Delta I/2}{J} \quad [\text{cm}^2]$$

$$A_{w(B)} = \frac{(5)(0.25)}{315}$$

$$A_{w(B)} = 0.0167 \quad [\text{cm}^2]$$

Step 6. Select a wire size from the wire table (Table 6.1), column 2. If the area is not within 10%, take the next smallest size. Also record micro-ohms per centimeter from column 4 and wire area with insulation A_w from column 5.

AWG No. 15

$$\text{Bare}, A_{w(B)} = 0.0165 \quad [\text{cm}^2]$$

$$\text{Insulated}, A_w = 0.0184 \quad [\text{cm}^2]$$

$$\mu\Omega/\text{cm} = 104$$

Step 7. Calculate the effective window area $W_{a(\text{eff})}$. Use window area W_a found in step 3:

$$W_{a(\text{eff})} = W_a S_3 \quad [\text{cm}^2]$$

A typical value for S_3 is 0.75, as shown in Chapter 6.

$$W_{a(\text{eff})} = (2.87)(0.75)$$

$$W_{a(\text{eff})} = 2.153 \quad [\text{cm}^2]$$

Step 8. Calculate the number of turns N. Use wire area A_w found in step 6:

$$N = \frac{W_{a(\text{eff})} S_2}{A_w} \quad [\text{turns}]$$

A typical value for S_2 is 0.6, as shown in Chapter 6.

$$N = \frac{(2.153)(0.6)}{0.0184}$$

$$N = 70 \quad [\text{turns}]$$

Step 9. Calculate the required gap, and use iron area A_c found in step 3:

$$l_g = \frac{0.4N^2A_c \times 10^{-8}}{L} \quad [\text{cm}]$$

$$l_g = \frac{1.26(70)^2(1.08) \times 10^{-8}}{0.002}$$

$$l_g = 0.0333 \quad [\text{cm}]$$

Gap spacing is usually maintained by inserting kraft paper. However, this paper is available only in mil thicknesses. Since l_g has been determined in centimeters, it is necessary to convert as follows:

$$\text{cm} \times 393.7 = \text{mils}$$

Substituting values:

$$(0.0333[\text{cm}])(393.7) = 13.1 \quad [\text{mils}]$$

Round off to nearest even mil and multiply:

$$(14 \times 10^{-3})(2.54) = 0.0355 \quad [\text{cm}]$$

Step 10. Calculate the amount of fringing flux. Use the *G* dimension found in step 3:

$$F = 1 + \frac{l_g}{A_c} \ln\left(\frac{2G}{l_g}\right)$$

$$F = 1 + \frac{0.0355}{1.04} \ln\left(\frac{2(3.016)}{0.0355}\right)$$

$$F = 1.175$$

Step 11. Calculate the new number of turns by inserting the fringing flux:

$$N = \frac{l_g L}{0.4\pi A_c F \times 10^{-8}} \quad [\text{turns}]$$

$$N = \frac{(0.0355)(0.002)}{(1.26)(1.08)(1.17) \times 10^{-8}}$$

$$N = 67 \quad [\text{turns}]$$

Step 12. Calculate the winding resistance. Use MLT from step 4 and micro-ohms per-centimeter from step 6.

$$R = \text{MLT} \times N \frac{\mu\Omega}{\text{cm}} \times 10^{-6} \quad [\Omega]$$

$$R = 6.6 \times 67 \times 104 \times 10^{-6}$$

$$R = .046 \quad [\Omega]$$

Step 13. Calculate the copper loss P_{cu}:

$$P_{cu} = I^2 R \quad [\text{W}]$$

$$P_{cu} = 5.25^2 \times .046$$

$$P_{cu} = 1.25 \quad [\text{W}]$$

Step 14. Calculate the total ac plus dc flux density:

$$B_m = \frac{0.4\pi N I_{dc} + \dfrac{\Delta I}{2} \times 10^{-4}}{l_g} \quad [\text{tesla}]$$

$$B_m = \frac{1.26(67)5 + \dfrac{0.5}{2} \times 10^{-4}}{(.0355)}$$

$$B_m = 1.24 \quad [\text{tesla}]$$

Step 15. Calculate the ac flux density:

$$B_m = \frac{0.4\pi N \dfrac{\Delta I}{2} = 10^{-4}}{l_g} \quad [\text{tesla}]$$

$$B_m = \frac{1.26 \times 67 \times 0.25 \times 10^{-4}}{.0355}$$

$$B_m = .0594 \quad [\text{tesla}]$$

Step 16. Calculate the watts per kilogram for the appropriate core material (see Chapter 3), then determine the core loss P_{fe}. Core weight is found in step 3:

$$\text{W/kg} = K f^{(m)} B_m^{(n)}$$

$$\text{W/kg} = (5.97)(20,000)^{1.26}(0.0594)^{1.73} \times 10^{-3}$$

$$\text{W/kg} = 11.8$$

$$P_{fe} = (\text{W/kg}) W_{tfe} \quad [\text{W}]$$

$$P_{fe} = (11.8)(0.088)$$

$$P_{fe} = 1.038 \quad [\text{W}]$$

Step 17. Calculate the total losses P_Σ:

$$P_\Sigma = P_{\text{cu}} + P_{\text{fe}} \quad [\text{W}]$$

$$P_\Sigma = 0.734 + 1.038$$

$$P_\Sigma = 1.77 \quad [\text{W}]$$

Step 18. Calculate the watts per unit area. The surface area A_t is found in step 3:

$$\psi = \frac{P_\Sigma}{A_t} \quad [\text{W/cm}^2]$$

$$\psi = \frac{1.77}{69.1}$$

$$\psi = 0.0256 \quad [\text{W/cm}^2]$$

4.7 GAPPED INDUCTOR DESIGN USING THE CORE GEOMETRY K_g APPROACH: AN EXAMPLE

Design work sheets for a gapped inductor with the following specifications:

1. Inductance, $L = 0.001$ [H]
2. dc current, $I_o = 5$ [A]
3. ac current, $\Delta I = 0.5$ [A]
4. Output power, $P_o = 100$ [W]
5. Regulation, $\alpha = 2$ [%]
6. Ripple frequency, $f = 20,000$ [Hz]
7. Flux density, $B_m = 1.4$ [T]
8. Core material: silicon
9. Core configuration: C core

Step 1. Calculate the energy-handling capability:

$$\text{Energy} = \frac{LI^2}{2} \quad [\text{W-s}]$$

$$I = I_o + \frac{\Delta I}{2} \quad [\text{A}]$$

$$I = 5 + \frac{0.5}{2}$$

$$I = 5.25 \quad [\text{A}]$$

$$\text{Energy} = \frac{(0.001)(5.25)^2}{2}$$

$$\text{Energy} = 0.0138 \quad [\text{W-s}]$$

Step 2.   Calculate the electrical conditions constant K_e:

$$K_e = 0.145 P_o B_m^2 \times 10^{-4}$$

$$K_e = 0.145(100)(1.4)^2 \times 10^{-4}$$

$$K_e = 0.00284$$

Step 3.   Calculate the core geometry coefficient K_g:

$$K_g = \frac{(\text{Energy})^2}{K_e \alpha} \quad [\text{cm}^5]$$

$$K_g = \frac{(0.0138)^2}{(0.00284)(2)}$$

$$K_g = 0.0335 \quad [\text{cm}^5]$$

Step 4.   Select a comparable core geometry coefficient K_g from the specified core section (see Chapter 3), and record the appropriate data:

Core CL-5

$$K_g = 0.0357 \quad [\text{cm}^5]$$

$$A_p = 0.76 \quad [\text{cm}^4]$$

$$\text{MLT} = 4.6 \quad [\text{cm}]$$

$$A_c = 0.54 \quad [\text{cm}^2]$$

$$W_a = 1.41 \quad [\text{cm}^2]$$

$$A_t = 33.7 \quad [\text{cm}^2]$$

$$G = 2.22 \quad [\text{cm}]$$

$$\text{MPL} = 7.4 \quad [\text{cm}]$$

$$W_{tfe} = 30 \quad [\text{g}]$$

Step 5. Calculate the current density J. Use the area product A_p found in step 4:

$$J = \frac{2(\text{Energy}) \times 10^4}{B_m A_p K_u} \quad [\text{A/cm}^2]$$

Window utilization factor $K_u = 0.4$

$$J = \frac{(2)(0.0138) \times 10^4}{(1.4)(0.76)(0.4)}$$

$$J = 648 \quad [\text{A/cm}^2]$$

Step 6. Calculate the bare wire size $A_{w(B)}$:

$$A_{w(B)} = \frac{I_o + \Delta I/2}{J} \quad [\text{cm}^2]$$

$$A_{w(B)} = \frac{5 + 0.25}{648}$$

$$A_{w(B)} = 0.00810 \quad [\text{cm}^2]$$

Step 7. Select a wire size from Table 6.1, column 2. If the area is not within 10%, take the next smallest size. Also record micro-ohms per centimeter from column 4 and wire area with insulation A_w from column 5.

AWG No. 18

Bare, $A_{w(B)} = 0.00823 \quad [\text{cm}^2]$

Insulated, $A_w = 0.00932 \quad [\text{cm}^2]$

$\mu\Omega/\text{cm} = 209$

Step 8. Calculate the effective window area $W_{a(\text{eff})}$. Use window area W_a found in step 4:

$$W_{a(\text{eff})} = W_a S_3 \quad [\text{cm}^2]$$

A typical value for S_3 is 0.75, as shown in Chapter 6.

$$W_{a(\text{eff})} = (1.41)(0.75)$$

$$W_{a(\text{eff})} = 1.05 \quad [\text{cm}^2]$$

Step 9. Calculate the number of turns. Use wire area A_w found in step 7:

$$N = \frac{W_{a(\text{eff})} S_2}{A_w} \quad [\text{turns}]$$

A typical value for S_2 is 0.6, as shown in Chapter 6.

$$N = \frac{(1.05)(0.6)}{0.00932}$$

$$N = 68 \quad [\text{turns}]$$

Step 10. Calculate the required gap and use the iron area A_c found in step 4:

$$l_g = \frac{0.4N^2A_c \times 10^{-8}}{L} \quad [cm]$$

$$l_g = \frac{(1.26)(68)^2(0.54) \times 10^{-8}}{0.001}$$

$$l_g = 0.0314 \quad [cm]$$

Gap spacing is usually maintained by inserting kraft paper. However, this paper is available only in 1 mil thickness. Since l_g has been determined in centimeters, it is necessary to convert as follows:

$$cm \times 393.7 = mils$$

Substituting values:

$$(0.0314[cm]) \times 393.7 = 12.4 \quad [mils]$$

Round off to nearest even mil and multiply:

$$(0.014)2.54 = 0.0355 \quad [cm]$$

Step 11. Calculate the amount of fringing flux. Use the G dimension found in step 4:

$$F = 1 + \frac{l_g}{A_c} \ln\left[\frac{2G}{l_g}\right]$$

$$F = 1 + \frac{0.0355}{0.76} \ln\left[\frac{2(2.22)}{0.0355}\right]$$

$$F = 1.23$$

Step 12. Calculate the new number of turns by inserting the fringing flux:

$$N = \frac{l_g L}{0.4\pi A_c F \times 10^{-8}} \quad \text{[turns]}$$

$$N = \frac{(0.355)(0.001)}{(1.26)(0.54)(1.23) \times 10^{-8}}$$

$$N = 64 \quad \text{[turns]}$$

Step 13.   Calculate the winding resistance. Use MLT from step 4 and micro-ohms per centimeter from step 7.

$$R = (\text{MLT})(N)\left(\frac{\mu\Omega}{\text{cm}}\right) \times 10^{-6} \quad [\Omega]$$

$$R = (4.6)(64)(209) \times 10^{-6}$$

$$R = 0.0615 \quad [\Omega]$$

Step 14.   Calculate the copper loss P_{cu}:

$$P_{cu} = I^2 R \quad \text{[W]}$$

$$P_{cu} = (5.25)^2 (0.0615)$$

$$P_{cu} = 1.69 \quad \text{[W]}$$

Step 15.   Calculate the regulation α:

$$\alpha = \frac{P_{cu}}{P_o} \times 100 \quad \text{[\%]}$$

$$\alpha = \frac{1.69 \times 100}{100}$$

$$\alpha = 1.70 \quad \text{[\%]}$$

Step 16. Calculate the total ac plus dc flux density:

$$B_m = \frac{0.4\pi N(I_{dc} + \Delta I/2) \times 10^{-4}}{(l_g)} \quad [T]$$

$$B_m = \frac{(1.26)(64)(5 + 0.5/2) \times 10^{-4}}{0.0355}$$

$$B_m = 1.19 \quad [T]$$

Step 17. Calculate the ac flux density:

$$B_{ac} = \frac{0.4\pi N(\Delta I/2) \times 10^{-4}}{l_g} \quad [T]$$

$$B_{ac} = \frac{(1.26)(64)(0.25) \times 10^{-4}}{0.0355}$$

$$B_{ac} = 0.0576 \quad [T]$$

Step 18. Calculate the watts per kilogram for the appropriate core material (see Chapter 3); then determine the core loss. Core weight is found in step 4:

$$W/kg = Kf^{(m)}B_{ac}^{(n)} \quad [T]$$

$$mW/g = (5.97)(20,000)^{1.26}(0.0576)^{1.73} \times 10^{-3}$$

$$mW/g = 11.24$$

$$P_{fe} = (mW/g)(W_{tfe}) \times 10^{-3} \quad [W]$$

$$P_{fe} = (11.24)(30) \times 10^{-3}$$

$$P_{fe} = 0.337 \quad [W]$$

Step 19.   Calculate the total losses P_Σ:

$$P_\Sigma = P_{cu} + P_{fe} \quad [W]$$

$$P_\Sigma = 1.69 + 0.337$$

$$P_\Sigma = 2.03 \quad [W]$$

Step 20.   Calculate the efficiency of the inductor:

$$\eta = \frac{P_o \times 100}{P_o + P_\Sigma} \quad [\%]$$

$$\eta = \frac{100 \times 100}{100 + 2.03}$$

$$\eta = 98 \quad [\%]$$

Step 21.   Calculate the watts per unit area. The surface area A_t is found in step 4:

$$\psi = \frac{P_\Sigma}{A_t} \quad [W/cm^2]$$

$$\psi = \frac{2.03}{33.7}$$

$$\psi = 0.0602 \quad [W/cm^2]$$

REFERENCES

1. *Molypermalloy Powder Cores.* Catalog MPP-3035, Magnetics Inc., Butler, Pennsylvania.

2. Lee, R., *Electronic Transformer and Circuits,* 2nd ed., John Wiley & Sons, New York, 1958.

3. *Silectron Cores,* Bulletin SC-107B, Arnold Engineering, Marengo, Illinois, undated.

4. *Orthosil Wound Cores,* Catalog No. W102-C, Thomas & Skinner, Inc., Indianapolis, Indiana, undated.

APPENDIX A. LINEAR REACTOR DESIGN WITH AN IRON CORE

After calculating the inductance and dc current, select the proper size core with a given $LI^2/2$. The energy-handling capability of an inductor can be determined by its area product $A_p W_a A_c$, where W_a is the available core window area in cm^2 and A_c is the core effective cross-sectional area in cm^2. The $W_a A_c$ or area product A_p relationship is obtained by solving $E = L \, dI/dt$ as follows*:

$$E = L \frac{dI}{dt} = N \frac{d\phi}{dt} \tag{4.A-1}$$

$$L = N \frac{d\phi}{dI} \tag{4.A-2}$$

$$\phi = B_m A_c' \tag{4.A-3}$$

$$B_m = \frac{\mu_0 NI}{l_g' + l_m'/\mu_r} \tag{4.A-4}$$

$$\phi = \frac{\mu_0 NI A_c'}{l_g' + l_m'/\mu_r} \tag{4.A-5}$$

$$\frac{d\phi}{dI} = \frac{\mu_0 N A_c'}{l_g' + l_m'/\mu_r} \tag{4.A-6}$$

$$L = N \frac{d\phi}{dI} = \frac{\mu_0 N^2 A_c'}{l_g' + l_m'/\mu_r} \tag{4.A-7}$$

$$\text{Energy} = \frac{1}{2} LI^2 = \frac{\mu_0 N^2 A_c' I^2}{2(l_g' + l_m'/\mu_r)} \tag{4.A-8}$$

If B_m is specified,

$$I = \frac{B_m(l_g' + l_m'/\mu_r)}{\mu_0 N} \tag{4.A-9}$$

---

*Symbols marked with a prime (such as H') are mks (meter-kilogram-second) units.

$$\text{Energy} = \frac{\mu_0 N^2 A_c'}{2(l_g' + l_m'/\mu_r)} \left(\frac{B_m(l_g' + l_m'/\mu_r)}{\mu_0 N} \right)^2 \qquad (4.A\text{-}10)$$

$$\text{Energy} = \frac{B_m^2(l_g' + l_m'/\mu_r)A_c'}{2\mu_0} \qquad (4.A\text{-}11)$$

$$I = \frac{K_u W_a' J'}{N} = \frac{B_m(l_g' + l_m'/\mu_r)}{\mu_0 N} \qquad (4.A\text{-}12)$$

Solving for $l_g' + l_m'/\mu_r$:

$$l_g' + \frac{l_m'}{\mu_r} = \frac{\mu_0 K_u W_a' J'}{B_m} \qquad (4.A\text{-}13)$$

Substituting into the energy equation:

$$\text{Energy} = \frac{B_m^2(\mu_0 K_u W_a' J'/B_m)A_c'}{2\mu_0} \qquad (4.A\text{-}14)$$

$$\text{Energy} = \left(\frac{B_m^2 A_c'}{2\mu_0} \right) \left(\frac{\mu_0 K_u W_a' J'}{B_m} \right) \qquad (4.A\text{-}15)$$

$$\text{Energy} = \frac{B_m K_u W_a' A_c' J'}{2} \qquad (4.A\text{-}16)$$

If W_a is the window area, cm^2
$\quad A_c$ is the core area, cm^2
$\quad J$ is the current density, A/cm^2
$\quad H$ is the magnetizing force, amp-turns/cm
$\quad l_g$ is the air gap, cm
$\quad l_m$ is the magnetic path length, cm

then

$$W_a' = W_a \times 10^{-4}$$
$$A_c' = A_c \times 10^{-4}$$
$$J' = J \times 10^4$$
$$l_m' = l_m \times 10^{-2}$$
$$l_g' = l_g \times 10^{-2}$$
$$H' = H \times 10^2$$

and we can substitute into the energy equation to obtain

$$\text{Energy} = \frac{W_a A_c B_m J K_u}{2} \times 10^{-4}$$

(4.A-17)

Solving for $A_p = W_a A_c$,

$$A_p = \frac{2(\text{Energy})}{B_m J K_u} \times 10^4$$

(4.A-18)

Using the data from Table 3-1

$$J = K_j A_p^{-0.14}$$

(4.A-19)

yielding

$$A_p = \frac{2(\text{Energy}) \times 10^4}{K_u B_m (K_j A_p^{-0.14})}$$

(4.A-20)

$$A_p^{0.86} = \frac{2(\text{Energy}) \times 10^4}{K_u B_m K_j}$$

(4.A-21)

$$A_p = \left(\frac{2(\text{Energy}) \times 10^4}{K_u B_m K_j} \right)^{1.16} \quad [\text{cm}^4]$$

(4.A-22)

APPENDIX B. INDUCTORS DESIGNED FOR A GIVEN REGULATION

Inductors, like transformers, are designed for a given temperature rise. They can also be designed for a given regulation. The regulation and energy-handling ability of a core is related to the two constants K_g and K_e:

$$(\text{Energy})^2 = K_g K_e \alpha$$

(4.B-1)

where α is the regulation, %.

The constant K_g is determined by the core geometry, it can be expressed as the function

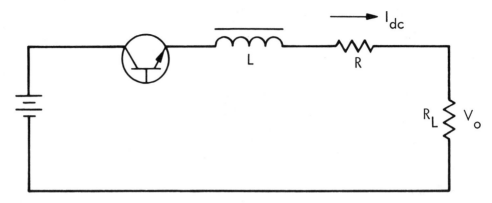

Figure 4.B-1. Output inductor.

$$K_g = f(A_c, W_a, \text{MLT}) \qquad (4.\text{B-2})$$

The constant K_e is determined by the magnetic and electric operating conditions, it can be expressed as

$$K_e = f(P_o, B_m)$$

For the circuit shown in Figure 4.B-1, the derivation of the specific functions for K_g and K_e is as follows:

$$P_o = I_{dc}V_o \quad [\text{W}] \qquad (4.\text{B-4})$$

$$\alpha = \frac{I_{dc}R}{V_o} \times 100 \quad [\%] \qquad (4.\text{B-5})$$

Inductance is equal to

$$L = \frac{0.4\pi N^2 A_c \times 10^{-8}}{l_g} \quad [\text{H}] \qquad (4.\text{B-6})$$

Flux density is equal to

$$B_{dc} = \frac{0.4\pi N I_{dc} \times 10^{-4}}{l_g} \quad [\text{T}] \qquad (4.\text{B-7})$$

Combining the two equations,

$$\frac{L}{B_{dc}} = \frac{N A_c \times 10^{-4}}{I_{dc}}$$

(4.B-8)

Solving for N,

$$N = \frac{L I_{dc} \times 10^4}{B_{dc} A_c}$$

(4.B-9)

Since the resistance equation is

$$R = \frac{\rho N^2 \, \text{MLT}}{K_u W_a} \quad [\Omega]$$

(4.B-10)

and the regulation equation is

$$\alpha = \frac{I_{dc} R}{V_o} \times 10^2 \quad [\%]$$

(4.B-11)

inserting the resistance Equation (4.B-10) gives

$$\alpha = \frac{I_{dc}}{V_o} \times \frac{\rho N^2 \, \text{MLT}}{K_u W_a} \times 10^2$$

(4.B-12)

$$N^2 = \left(\frac{L I_{dc}}{B_{dc} A_c}\right)^2 \times 10^8$$

(4.B-13)

$$\alpha = \frac{I_{dc} \, \text{MLT} \, \rho}{V_o I_u W_a} \times \left(\frac{L I_{dc}}{B_{dc} A_c}\right)^2 \times 10^{10}$$

(4.B-14)

$$\alpha = \frac{I_{dc} \, \text{MLT} \, \rho (L I_{dc})^2}{V_o K_u W_a B_{dc}^2 A_c^2} \times 10^{10}$$

(4.B-15)

$$\text{Energy} = \frac{L I_{dc}^2}{2} \quad [\text{W-s}]$$

(4.B-16)

Multiplying the equation by I_{dc}/I_{dc} and combining,

$$\alpha = \frac{(LI_{dc}^2)^2 \rho \ \text{MLT} \times 10^{10}}{V_o I_{dc} K_u W_a A_c^2 B_{dc}^2} \qquad \qquad (4.\text{B-}17)$$

which reduces to

$$\alpha = \frac{(2 \ \text{Energy})^2}{P_o B_{dc}^2} \times \frac{\rho \ \text{MLT}}{K_u W_a A_c^2} \times 10^{10} \qquad \qquad (4.\text{B-}18)$$

$$\rho = 1.724 \times 10^{-6} \quad [\Omega\text{-cm}]$$

$$\alpha = \frac{6.89 \ (\text{Energy})^2}{P_o B_{dc}^2} \times \frac{\text{MLT}}{K_u W_a A_c^2} \times 10^4 \qquad \qquad (4.\text{B-}19)$$

Solving for energy,

$$(\text{Energy})^2 = 0.145 P_o B_{dc}^2 \times \frac{K_u W_a A_c^2}{\text{MLT}} \times 10^{-4} \alpha \qquad \qquad (4.\text{B-}20)$$

$$K_g = \frac{K_u W_a A_c^2}{\text{MLT}} \quad [\text{cm}^5] \qquad \qquad (4.\text{B-}21)$$

$$K_e = 0.145 P_o B_{dc}^2 \times 10^{-4} \qquad \qquad (4.\text{B-}22)$$

$$\alpha = \frac{P_{cu}}{P_o} \times 100 \quad [\%] \qquad \qquad (4.\text{B-}23)$$

The regulation of an inductor is related to the copper loss, as shown in Equation 4.B-24.

$$\alpha = \frac{P_{cu}}{P_o} \times 100 \quad [\%] \qquad \qquad (4.\text{B-}24)$$

The copper loss in an inductor is related to the rms current. The rms current in a down regulator, as shown in Figure 4.B-1, is always equal to or less than I_o:

$$I_{rms} \leqslant I_o \qquad \qquad (4.\text{B-}25)$$

APPENDIX C. MAGNETIC AND DIMENSIONAL SPECIFICATIONS OF C CORES AND BOBBINS

C.1. Definitions for Tables 4.C-1 Through 4.C-20

Tables 4.C-1–4.C-20 show magnetic and dimensional specifications for 20 C cores. The information is listed by line as:

1	Manufacture and part number
2	Units
3	Ratio of the window area to the iron area, W_a/A_c
4	Product of the window area times the iron area
5	Window area W_a gross
6	Iron area A_c effective
7	Mean magnetic path length l_m
8	Core weight of silicon steel multiplied by the stacking factor
9	Copper weight, single bobbin
10	Mean length turn
11	Ratio of G dimension divided by the square root of the iron area A_c
12	Ratio of the $W_a(eff)/W_a$
13	Inductor overall surface area A_t
14–17	C-core dimensions
18	Bobbin manufacturer and part number* †
19	Bobbin inside winding length †
20	Bobbin inside build †
21	Bobbin winding area length times build †
22	Bracket manufacturer and part number ‡

C.2. Nomographs for 20 C Core Sizes

Figures 4.C-1–4.C-20 are graphs for 20 different C cores. The nomographs display resistance, number of turns, and wire size at a fill factor of $K_2 = 0.60$. These graphs are included to provide close approximations for breadboarding purposes. Figure 4.C-21 is a graph for inductance, capacitance, and reactance, and Figure 4.C-22 is a graph showing the area product versus the energy capability of C cores.

---

*The first number in front of the part number indicates the number of bobbins.
† Dorco Electronics, 15533 Vermont Ave., Paramount, California 90723.
‡ Hallmark Metals, 610 West Foothill Blvd., Glendora, California 91740.

Table 4.C-1. C core AL-2

"C" CORE	AL-2			
	ENGLISH		METRIC	
Wa/Ac			3.32	
Wa x Ac	0.0073	in^4	0.265	cm^4
Wa	0.156	in^2	1.006	cm^2
Ac (effective)	0.041	in^2	0.264	cm^2
lm	2.233	in	5.671	cm
CORE WT	0.027	lb	12.23	grams
COPPER WT	0.371	lb	16.87	grams
* MLT FULLWOUND	1.76	in	4.47	cm
G/√Ac			3.08	
Wa (effective) /Wa			0.835	
A$_T$	3.80	in^2	24.56	cm^2
D	0.250	in	0.635	cm
E	0.187	in	0.474	cm
F	0.250	in	0.635	cm
G	0.625	in	1.587	cm
BOBBIN	DORCO ELECTRONICS #		1-L-2	
LENGTH	0.580	in	1.473	cm
BUILD	0.225	in	0.571	cm
* Wa (effective)	0.130	in^2	0.841	cm^2
BRACKET	HALLMARK METALS #		04-010-03	

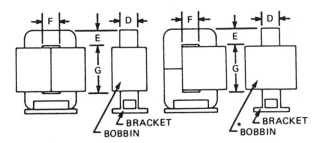

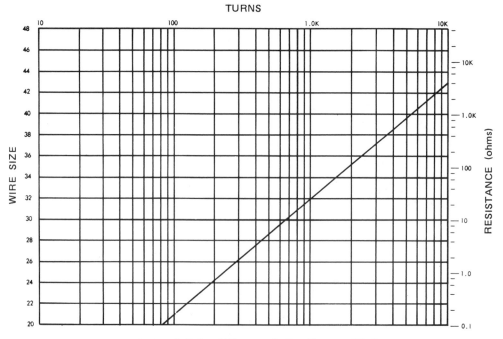

Figure 4.C-1. Wiregraph for C core AL-2.

Table 4.C-2. C core AL-3

"C" CORE	AL-3			
	ENGLISH		METRIC	
Wa/Ac			2.23	
Wa x Ac	0.0098	in^4	0.410	cm^4
Wa	0.156	in^2	1.006	cm^2
Ac (effective)	0.063	in^2	0.406	cm^2
lm	2.233	in	5.671	cm
CORE WT	0.04	lb	18.12	grams
COPPER WT	0.042	lb	19.25	grams
* MLT FULLWOUND	2.01	in	5.10	cm
G/√Ac			2.49	
Wa (effective) /Wa			0.835	
A$_T$	4.27	in^2	27.58	cm^2
D	0.375	in	0.952	cm
E	0.187	in	0.474	cm
F	0.250	in	0.635	cm
G	0.625	in	1.587	cm
BOBBIN	DORCO ELECTRONICS #	1-L-3		
LENGTH	0.580	in	1.473	cm
BUILD	0.225	in	0.571	cm
* Wa (effective)	0.130	in^2	0.841	cm^2
BRACKET	HALLMARK METALS #	06-010-03		

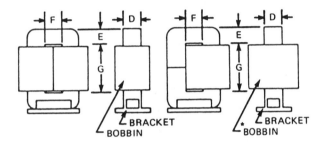

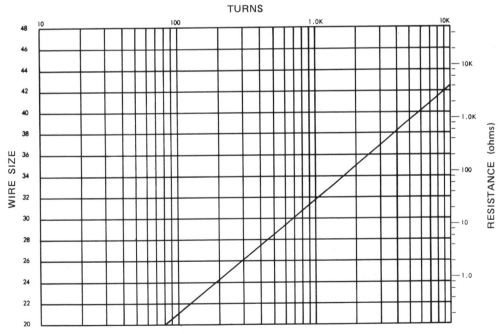

Figure 4.C-2. Wiregraph for C core AL-3.

Table 4.C-3. C core AL-5

"C" CORE	AL-5			
	ENGLISH		METRIC	
Wa/Ac			2.33	
Wa x Ac	0.018	in^4	0.767	cm^4
Wa	0.219	in^2	1.423	cm^2
Ac (effective)	0.0836	in^2	0.539	cm^2
lm	2.933	in	7.45	cm
CORE WT	0.067	lb	30.4	grams
COPPER WT	0.0643	lb	29.2	grams
* MLT FULLWOUND	2.13	in	5.42	cm
G/√Ac			3.026	
Wa (effective) /Wa			0.843	
A$_T$	5.90	in^2	38.1	cm^2
D	0.375	in	0.952	cm
E	0.250	in	0.635	cm
F	0.250	in	0.635	cm
G	0.875	in	2.22	cm
BOBBIN	DORCO ELECTRONICS #		1-L-5	
LENGTH	0.830	in	2.11	cm
BUILD	0.225	in	0.571	cm
* Wa (effective)	0.186	in^2	1.20	cm^2
BRACKET	HALLMARK METALS #		06-012-04	

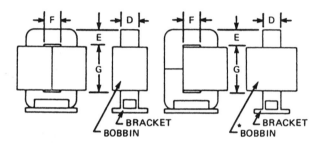

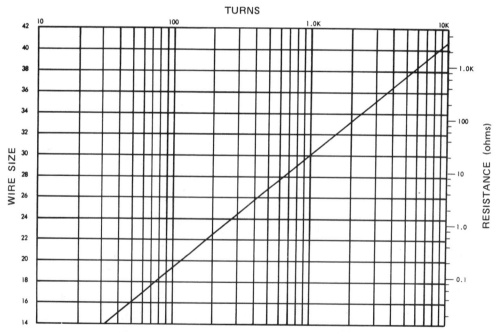

Figure 4.C-3. Wiregraph for C core AL-5.

Table 4.C-4. C core AL-6

"C" CORE	AL-6			
	ENGLISH		METRIC	
Wa/Ac			1.75	
Wa x Ac	0.024	in^4	1.011	cm^4
Wa	0.219	in^2	1.413	cm^2
Ac (effective)	0.111	in^2	0.716	cm^2
lm	2.933	in	7.45	cm
CORE WT	0.091	lb	41.2	grams
COPPER WT	0.0719	lb	32.6	grams
* MLT FULLWOUND	2.38	in	6.06	cm
G/√Ac			2.63	
Wa (effective) /Wa			0.843	
A$_T$	6.50	in^2	41.9	cm^2
D	0.500	in	1.27	cm
E	0.250	in	0.635	cm
F	0.250	in	0.635	cm
G	0.875	in	2.22	cm
BOBBIN	DORCO ELECTRONICS #	1-L-6		
LENGTH	0.830	in	2.11	cm
BUILD	0.225	in	0.571	cm
* Wa (effective)	0.186	in^2	1.20	cm^2
BRACKET	HALLMARK METALS #	08-012-04		

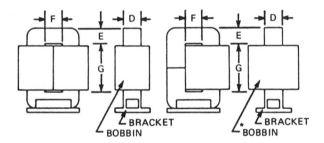

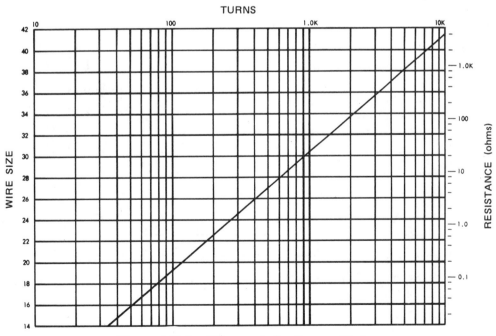

Figure 4.C-4. Wiregraph for C core AL-6.

Table 4.C-5. C core AL-124

"C" CORE	AL-124			
	ENGLISH		METRIC	
Wa/Ac			2.50	
Wa x Ac	0.0347	in^4	1.44	cm^4
Wa	0.313	in^2	2.02	cm^2
Ac (effective)	0.111	in^2	0.716	cm^2
lm	3.308	in	8.40	cm
CORE WT	0.103	lb	46.7	grams
COPPER WT	0.115	lb	52.13	grams
* MLT FULLWOUND	2.58	in	6.56	cm
G/√Ac			3.00	
Wa (effective) /Wa			0.876	
A$_T$	8.03	in^2	51.79	cm^2
D	0.500	in	1.27	cm
E	0.250	in	0.635	cm
F	0.313	in	0.795	cm
G	1.00	in	2.54	cm
BOBBIN	DORCO ELECTRONICS #		1-L-124	
LENGTH	0.955	in	2.425	cm
BUILD	0.288	in	0.731	cm
* Wa (effective)	0.275	in^2	1.77	cm^2
BRACKET	HALLMARK METALS #		08-013-04	

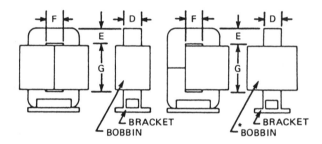

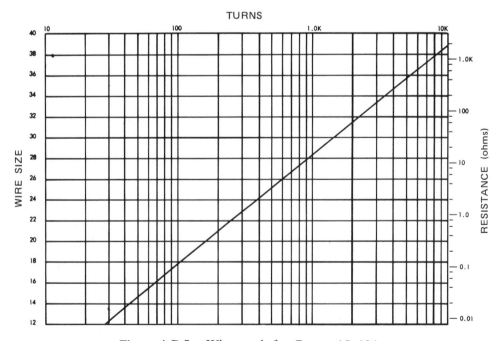

Figure 4.C-5. Wiregraph for C core AL-124.

Table 4.C-6. C core AL-8

"C" CORE	AL-8			
	ENGLISH		METRIC	
Wa/Ac			3.16	
Wa x Ac	0.056	in^4	2.31	cm^4
Wa	0.445	in^2	2.87	cm^2
Ac (effective)	0.125	in^2	0.806	cm^2
lm	4.198	in	10.66	cm
CORE WT	0.147	lb	66.59	grams
COPPER WT	0.180	lb	81.7	grams
* MLT FULLWOUND	2.77	in	7.06	cm
$G/\sqrt{Ac}$			3.36	
Wa (effective) /Wa			0.898	
A_T	11.29	in^2	72.8	cm^2
D	0.375	in	0.952	cm
E	0.375	in	0.952	cm
F	0.375	in	0.952	cm
G	1.187	in	3.015	cm
BOBBIN	DORCO ELECTRONICS #		1-L-8	
LENGTH	1.142	in	2.9	cm
BUILD	0.350	in	0.889	cm
* Wa (effective)	0.399	in^2	2.578	cm^2
BRACKET	HALLMARK METALS #		06-102-06	

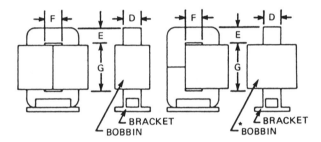

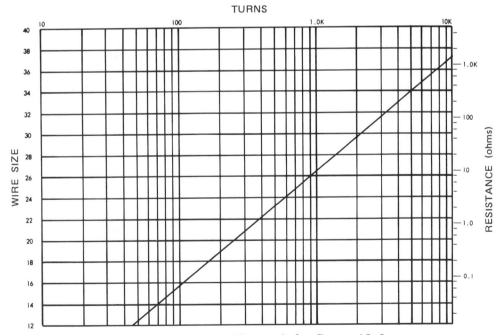

Figure 4.C-6. Wiregraph for C core AL-8.

Table 4.C-7. C core AL-9

"C" CORE	AL-9			
	ENGLISH		METRIC	
Wa/Ac			2.37	
Wa x Ac	0.074	in^4	3.09	cm^4
Wa	0.445	in^2	2.870	cm^2
Ac (effective)	0.167	in^2	1.077	cm^2
lm	4.198	in	10.66	cm
CORE WT	0.197	lb	89.2	grams
COPPER WT	0.196	lb	89.0	grams
* MLT FULLWOUND	3.02	in	7.69	cm
G/√Ac			2.90	
Wa (effective) /Wa			0.898	
A$_T$	12.15	in^2	78.39	cm^2
D	0.500	in	1.27	cm
E	0.375	in	0.952	cm
F	0.375	in	0.952	cm
G	1.187	in	3.015	cm
BOBBIN	DORCO ELECTRONICS #		1-L-9	
LENGTH	1.142	in	2.90	cm
BUILD	0.350	in	0.889	cm
* Wa (effective)	0.399	in^2	2.578	cm^2
BRACKET	HALLMARK METALS #		08-102-06	

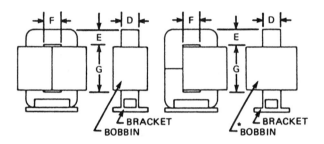

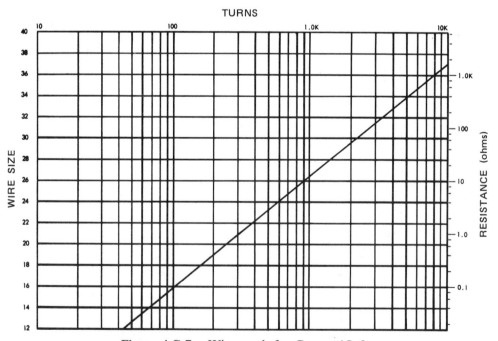

Figure 4.C-7. Wiregraph for C core AL-9.

Table 4.C-8. C core AL-10

"C" CORE	AL-10			
	ENGLISH		METRIC	
Wa/Ac			1.90	
Wa x Ac	0.092	in^4	3.85	cm^4
Wa	0.445	in^2	2.870	cm^2
Ac (effective)	0.208	in^2	1.342	cm^2
lm	4.198	in	10.66	cm
CORE WT	0.243	lb	110	grams
COPPER WT	0.213	lb	96.4	grams
* MLT FULLWOUND	3.27	in	8.33	cm
G/√Ac			2.603	
Wa (effective) /Wa			0.898	
A$_T$	13.01	in^2	83.9	cm^2
D	0.625	in	1.587	cm
E	0.375	in	0.952	cm
F	0.375	in	0.952	cm
G	1.187	in	3.015	cm
BOBBIN	DORCO ELECTRONICS #		1-L-10	
LENGTH	1.142	in	2.90	cm
BUILD	0.350	in	0.889	cm
* Wa (effective)	0.399	in^2	2.578	cm^2
BRACKET	HALLMARK METALS #		010-102-06	

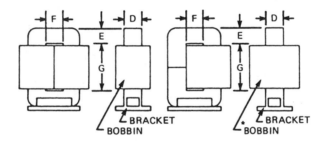

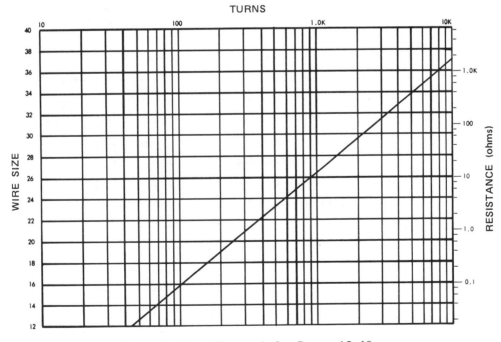

Figure 4.C-8. Wiregraph for C core AL-10.

Table 4.C-9. C core AL-12

"C" CORE	AL-12			
	ENGLISH		METRIC	
Wa/Ac			2.57	
Wa x Ac	0.109	in^4	4.57	cm^4
Wa	0.563	in^2	3.63	cm^2
Ac (effective)	0.195	in^2	1.26	cm^2
lm	4.523	in	11.5	cm
CORE WT	0.244	lb	110	grams
COPPER WT	0.295	lb	133.7	grams
* MLT FULLWOUND	3.54	in	9.00	cm
G/√Ac			2.55	
Wa (effective) /Wa			0.911	
A_T	15.61	in^2	100.7	cm^2
D	0.500	in	1.27	cm
E	0.437	in	1.11	cm
F	0.500	in	1.27	cm
G	1.125	in	2.857	cm
BOBBIN	DORCO ELECTRONICS #		1-L-12	
LENGTH	1.08	in	2.74	cm
BUILD	0.475	in	1.21	cm
* Wa (effective)	0.513	in^2	3.31	cm^2
BRACKET	HALLMARK METALS #		08-106-07	

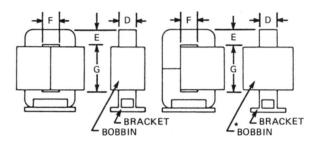

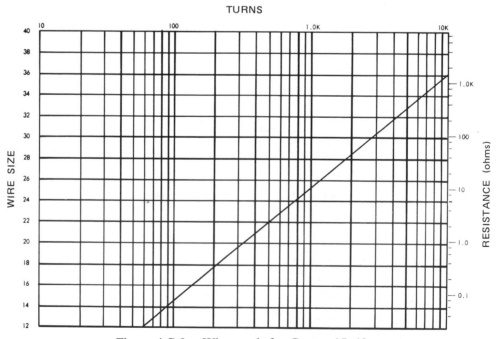

Figure 4.C-9. Wiregraph for C core AL-12.

Table 4.C-10. C core AL-135

"C" CORE	ENGLISH		METRIC	
AL-135				
Wa/Ac			2.89	
Wa x Ac	0.123	in^4	5.14	cm^4
Wa	0.633	in^2	4.083	cm^2
Ac (effective)	0.195	in^2	1.26	cm^2
lm	4.648	in	11.8	cm
CORE WT	0.251	lb	114	grams
COPPER WT	0.312	lb	159	grams
* MLT FULLWOUND	3.74	in	9.50	cm
G/√Ac			2.55	
Wa (effective) /Wa			0.915	
A$_T$	17.04	in^2	110	cm^2
D	0.500	in	1.27	cm
E	0.437	in	1.11	cm
F	0.562	in	1.43	cm
G	1.125	in	2.857	cm
BOBBIN	DORCO ELECTRONICS #		1-L-135	
LENGTH	1.08	in	2.74	cm
BUILD	0.537	in	1.36	cm
* Wa (effective)	0.579	in^2	3.74	cm^2
BRACKET	HALLMARK METALS #		08-107-07	

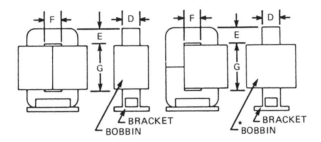

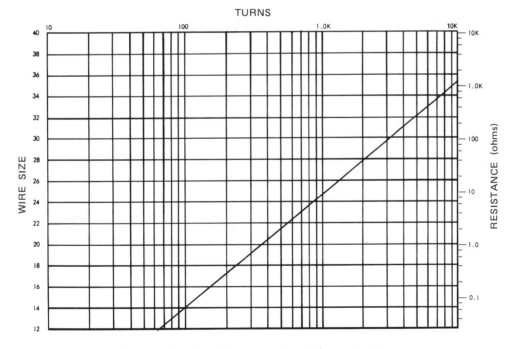

Figure 4.C-10. Wiregraph for C core AL-135.

Table 4.C-11.   C core AL-78

"C" CORE	ENGLISH		METRIC	
Wa/Ac			3.00	
Wa x Ac	0.146	in⁴	6.07	cm⁴
Wa	0.703	in²	4.53	cm²
Ac (effective)	0.208	in²	1.34	cm²
lm	5.891	in	14.96	cm
CORE WT	0.342	lb	154	grams
COPPER WT	0.331	lb	150	grams
* MLT FULLWOUND	3.21	in	8.15	cm
G/√Ac			4.93	
Wa (effective) /Wa			0.905	
A$_T$	16.99	in²	109.6	cm²
D	0.750	in	1.91	cm
E	0.313	in	0.795	cm
F	0.313	in	0.795	cm
G	2.250	in	5.715	cm
BOBBIN	DORCO ELECTRONICS #		1-L-78	
LENGTH	2.205	in	5.60	cm
BUILD	0.288	in	0.731	cm
* Wa (effective)	0.635	in²	4.10	cm²
BRACKET	HALLMARK METALS #		012-015-05	

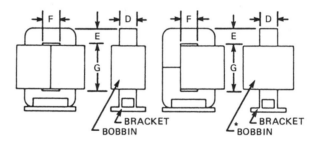

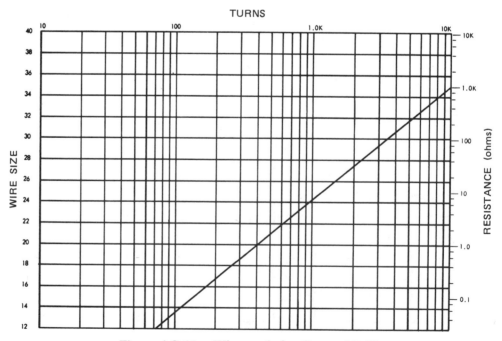

Figure 4.C-11.   Wiregraph for C core AL-78.

Table 4.C-12. C core AL-18

"C" CORE	AL-18			
	ENGLISH		METRIC	
Wa/Ac			5.08	
Wa x Ac	0.189	in^4	7.87	cm^4
Wa	0.977	in^2	6.30	cm^2
Ac (effective)	0.194	in^2	1.257	cm^2
lm	5.648	in	14.34	cm
CORE WT	0.305	lb	138	grams
COPPER WT	0.575	lb	260	grams
* MLT FULLWOUND	2.95	in	7.51	cm
G/$\sqrt{Ac}$			3.502	
Wa (effective) /Wa			0.890	
A$_T$	21.93	in^2	141.50	cm^2
D	0.500	in	1.27	cm
E	0.437	in	1.111	cm
F	0.625	in	1.587	cm
G	1.562	in	3.927	cm
BOBBIN	DORCO ELECTRONICS #		1-L-18	
LENGTH	1.497	in	3.802	cm
BUILD	0.590	in	1.498	cm
* Wa (effective)	0.880	in^2	5.697	cm^2
BRACKET	HALLMARK METALS #		08-108-07	

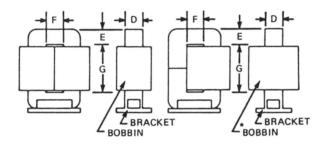

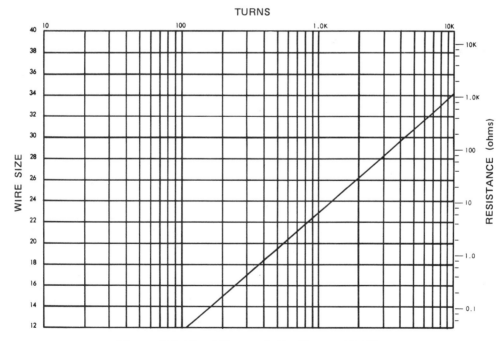

Figure 4.C-12. Wiregraph for C core AL-18.

Table 4.C-13. C core AL-15

"C" CORE	AL-15			
	ENGLISH		METRIC	
Wa/Ac			2.50	
Wa x Ac	0.218	in^4	9.07	cm^4
Wa	0.781	in^2	5.037	cm^2
Ac (effective)	0.279	in^2	1.80	cm^2
lm	5.588	in	14.2	cm
CORE WT	0.436	lb	197	grams
COPPER WT	0.448	lb	203	grams
* MLT FULLWOUND	3.97	in	10.08	cm
G/√Ac			2.96	
Wa (effective) /Wa			0.891	
A$_T$	21.07	in^2	135.9	cm^2
D	0.625	in	1.587	cm
E	0.500	in	1.27	cm
F	0.500	in	1.27	cm
G	1.562	in	3.967	cm
BOBBIN	DORCO ELECTRONICS #		1-L-15	
LENGTH	1.497	in	3.80	cm
BUILD	0.465	in	1.18	cm
* Wa (effective)	0.696	in^2	4.49	cm^2
BRACKET	HALLMARK METALS #		010-108-08	

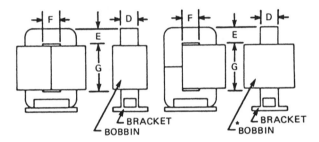

TURNS

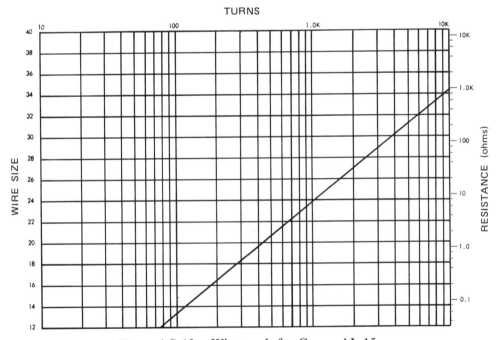

Figure 4.C-13. Wiregraph for C core AL-15.

Table 4.C-14. C core AL-16

"C" CORE	AL-16			
	ENGLISH		METRIC	
Wa/Ac			2.08	
Wa x Ac	0.26	in^4	10.8	cm^4
Wa	0.781	in^2	5.037	cm^2
Ac (effective)	0.334	in^2	2.15	cm^2
lm	5.588	in	14.2	cm
CORE WT	0.519	lb	235	grams
COPPER WT	0.476	lb	216	grams
* MLT FULLWOUND	4.22	in	10.72	cm
G/$\sqrt{Ac}$			2.70	
Wa (effective) /Wa			0.891	
A_T	22.21	in^2	143.3	cm^2
D	0.750	in	1.905	cm
E	0.500	in	1.27	cm
F	0.500	in	1.27	cm
G	1.562	in	3.967	cm
BOBBIN	DORCO ELECTRONICS # 1-L-16			
LENGTH	1.497	in	3.80	cm
BUILD	0.465	in	1.18	cm
* Wa (effective)	0.696	in^2	4.49	cm^2
BRACKET	HALLMARK METALS # 012-108-08			

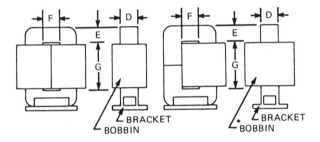

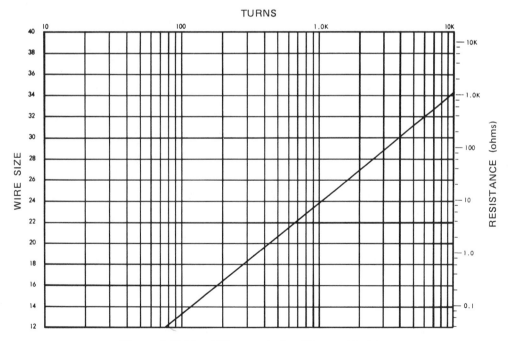

Figure 4.C-14. Wiregraph for C core AL-16.

Table 4.C-15.   C core AL-17

"C" CORE	AL-17			
	ENGLISH		METRIC	
Wa/Ac			1.56	
Wa x Ac	0.35	in^4	14.4	cm^4
Wa	0.781	in^2	5.037	cm^2
Ac (effective)	0.445	in^2	2.870	cm^2
lm	5.588	in	14.2	cm
CORE WT	0.693	lb	314	grams
COPPER WT	0.533	lb	241	grams
* MLT FULLWOUND	4.72	in	11.99	cm
G/√Ac			2.342	
Wa (effective) /Wa			0.891	
A$_T$	24.5	in^2	158	cm^2
D	1.000	in	2.54	cm
E	0.500	in	1.27	cm
F	0.500	in	1.27	cm
G	1.562	in	3.967	cm
BOBBIN	DORCO ELECTRONICS # 1-L-17			
LENGTH	1.497	in	3.80	cm
BUILD	0.465	in	1.18	cm
* Wa (effective)	0.696	in^2	4.49	cm^2
BRACKET	HALLMARK METALS # 10-108-08			

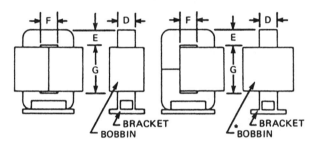

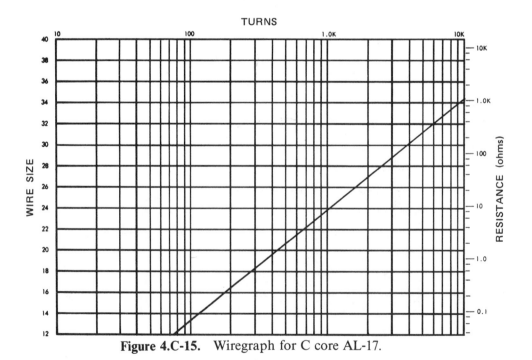

Figure 4.C-15.   Wiregraph for C core AL-17.

Table 4.C-16. C core AL-19

"C" CORE	AL-19			
	ENGLISH		METRIC	
Wa/Ac			1.95	
Wa x Ac	0.435	in^4	18.1	cm^4
Wa	0.977	in^2	6.30	cm^2
Ac (effective)	0.445	in^2	2.87	cm^2
lm	5.838	in	14.8	cm
CORE WT	0.724	lb	328	grams
COPPER WT	0.731	lb	332	grams
* MLT FULLWOUND	5.11	in	12.98	cm
G/√Ac			2.34	
Wa (effective) /Wa			0.903	
A$_T$	28.2	in^2	182	cm^2
D	1.000	in	2.54	cm
E	0.500	in	1.27	cm
F	0.625	in	1.587	cm
G	1.562	in	3.967	cm
BOBBIN	DORCO ELECTRONICS # 1-L-19			
LENGTH	1.497	in	3.80	cm
BUILD	0.590	in	1.498	cm
* Wa (effective)	0.883	in^2	5.69	cm^2
BRACKET	HALLMARK METALS # 10-110-08			

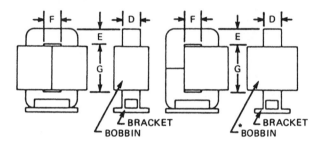

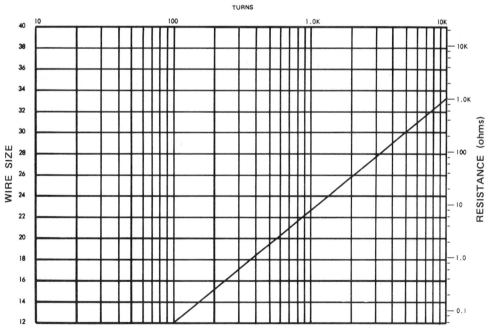

Figure 4.C-16. Wiregraph for C core AL-19.

Table 4.C-17. C core AL-20

"C" CORE	AL-20			
	ENGLISH		METRIC	
Wa/Ac			1.56	
Wa x Ac	0.543	in^4	22.6	cm^4
Wa	0.977	in^2	6.30	cm^2
Ac (effective)	0.556	in^2	3.58	cm^2
lm	6.228	in	15.8	cm
CORE WT	0.965	lb	437	grams
COPPER WT	0.767	lb	348	grams
* MLT FULLWOUND	5.36	in	13.62	cm
G/√Ac			2.09	
Wa (effective) /Wa			0.903	
A$_T$	31.7	in^2	205	cm^2
D	1.000	in	2.54	cm
E	0.625	in	1.587	cm
F	0.625	in	1.587	cm
G	1.562	in	3.967	cm
BOBBIN	DORCO ELECTRONICS # 1-L-20			
LENGTH	1.497	in	3.80	cm
BUILD	0.590	in	1.498	cm
* Wa (effective)	0.883	in^2	5.69	cm^2
BRACKET	HALLMARK METALS # 10-114-010			

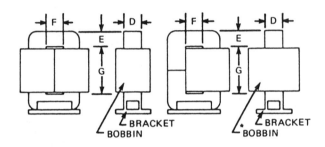

TURNS

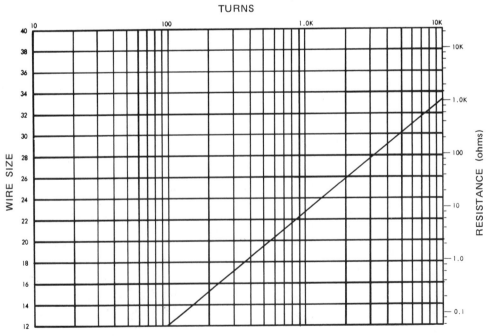

Figure 4.C-17. Wiregraph for C core AL-20.

Table 4.C-18. C core AL-22

"C" CORE	AL-22			
	ENGLISH		METRIC	
Wa/Ac			1.94	
Wa x Ac	0.692	in^4	28.0	cm^4
Wa	1.21	in^2	7.804	cm^2
Ac (effective)	0.556	in^2	3.58	cm^2
lm	6.978	in	17.2	cm
CORE WT	1.08	lb	489	grams
COPPER WT	0.961	lb	435	grams
* MLT FULLWOUND	5.36	in	13.62	cm
G/√Ac			2.598	
Wa (effective) /Wa			0.912	
A$_T$	35.3	in^2	228	cm^2
D	1.000	in	2.54	cm
E	0.625	in	1.587	cm
F	0.625	in	1.587	cm
G	1.937	in	4.92	cm
BOBBIN	DORCO ELECTRONICS #		1-L-22	
LENGTH	1.872	in	4.75	cm
BUILD	0.590	in	1.498	cm
* Wa (effective)	1.10	in^2	7.12	cm^2
BRACKET	HALLMARK METALS #		10-114-010	

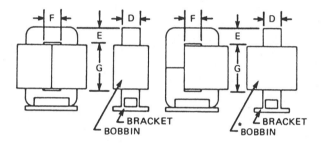

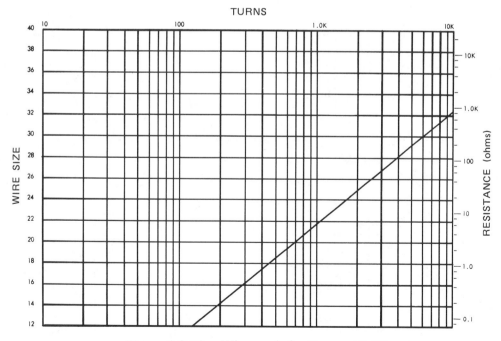

Figure 4.C-18. Wiregraph for C core AL-22.

Table 4.C-19. C core AL-23

"C" CORE		ENGLISH		METRIC	
Wa/Ac				1.55	
Wa x Ac		0.841	in⁴	34.96	cm⁴
Wa		1.21	in²	7.804	cm²
Ac (effective)		0.695	in²	4.48	cm²
lm		6.978	in	17.2	cm
CORE WT		1.352	lb	612	grams
COPPER WT		1.056	lb	479	grams
* MLT FULLWOUND		5.86	in	14.89	cm
G/√Ac				2.32	
Wa (effective) /Wa				0.912	
A_T		38.1	in²	246	cm²
D		1.250	in	3.175	cm
E		0.625	in	1.587	cm
F		0.625	in	1.587	cm
G		1.937	in	4.92	cm
BOBBIN		DORCO ELECTRONICS #	1-L-23		
LENGTH		1.872	in	4.75	cm
BUILD		0.590	in	1.498	cm
* Wa (effective)		1.10	in²	7.12	cm²
BRACKET		HALLMARK METALS #	14-114-010		

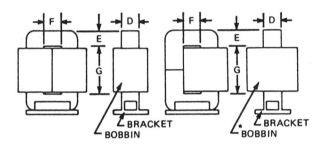

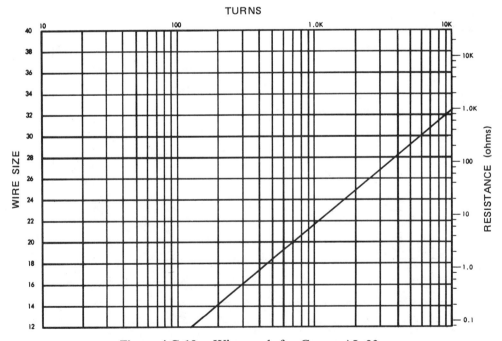

Figure 4.C-19. Wiregraph for C core AL-23.

Table 4.C-20.   C core AL-24

"C" CORE	AL-24			
	ENGLISH		METRIC	
Wa/Ac			2.77	
Wa x Ac	0.962	in^4	40.0	cm^4
Wa	1.73	in^2	11.16	cm^2
Ac (effective)	0.556	in^2	3.58	cm^2
lm	7.871	in	20.0	cm
CORE WT	1.220	lb	553	grams
COPPER WT	1.501	lb	680	grams
* MLT FULLWOUND	5.75	in	14.62	cm
G/√Ac			3.10	
Wa (effective) /Wa			0.929	
A$_T$	43.6	in^2	281.6	cm^2
D	1.000	in	2.54	cm
E	0.625	in	1.587	cm
F	0.750	in	1.905	cm
G	2.313	in	5.875	cm
BOBBIN	DORCO ELECTRONICS #		1-L-24	
LENGTH	2.248	in	5.709	cm
BUILD	0.715	in	1.816	cm
* Wa (effective)	1.607	in^2	10.37	cm^2
BRACKET	HALLMARK METALS #		10-200-010	

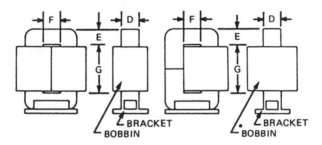

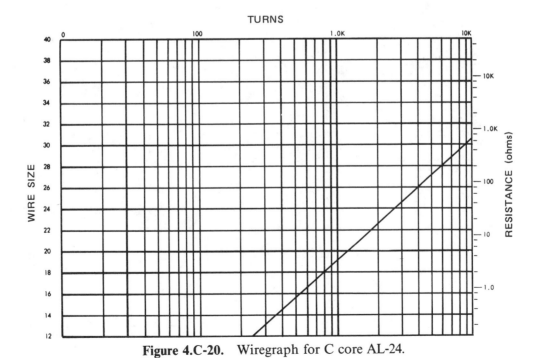

Figure 4.C-20.   Wiregraph for C core AL-24.

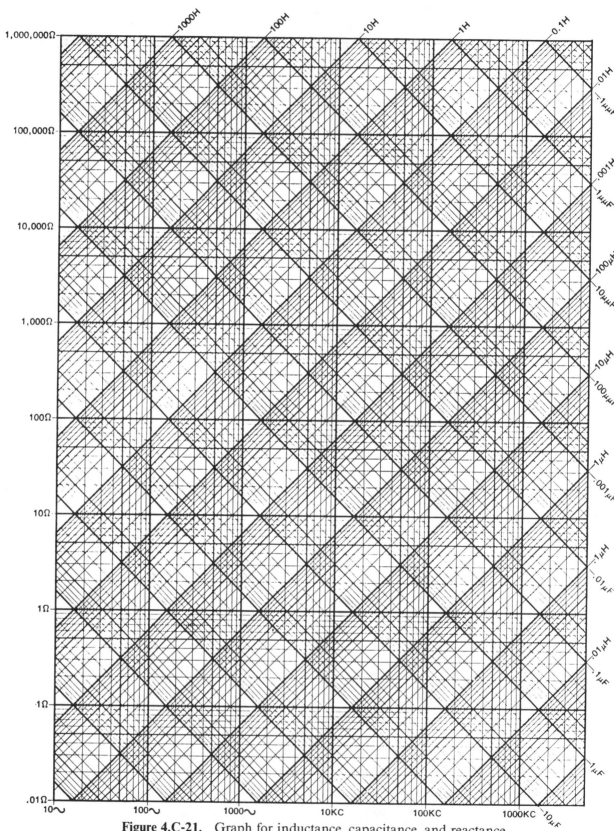

Figure 4.C-21. Graph for inductance, capacitance, and reactance.

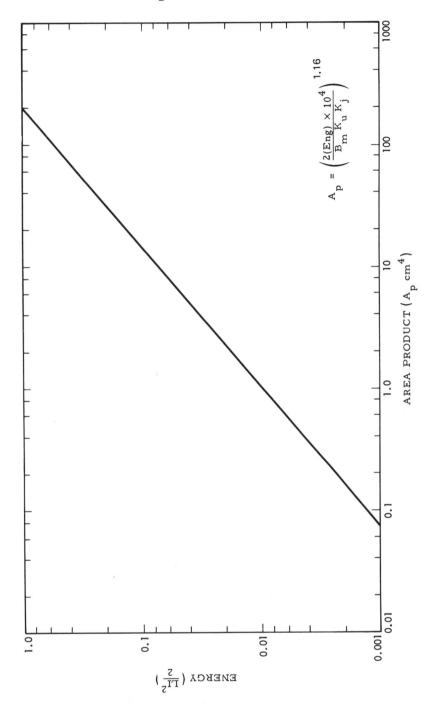

$$A_p = \left(\frac{2(\text{Eng}) \times 10^4}{B_m K_u K_j} \right)^{1.16}$$

Figure 4.C-22. Area product A_p vs. energy $LI^2/2$. $B_m = 1.2$ T; $K_u = 0.4$; $K_j = 395$.

Chapter 5

Toroidal Powder Core Selection With DC Current

5.1 INTRODUCTION*

5.1.1 Molybdenum Permalloy Powder Cores

Molybdenum permalloy powder cores (MPP cores) are manufactured from very fine particles of an 81% nickel, 17% iron, 2% molybdenum alloy. This powder is coated with an inert insulation to minimize eddy current losses and to introduce a distributed air gap into the core structure. The insulated powder is then compacted into toroidal cores ranging in size from 0.1 to 5 inches in outside diameter. MPP cores are available in permeabilities ranging from 14 up to 550.

MPP cores have unique magnetic properties that make them suitable for a wide range of applications. Some of the more important magnetic properties are:

1. Relatively high saturation flux density (0.7 T)
2. Low residual flux density (less than 0.050 T)
3. Very flat permeability versus temperature (less than 3% change in μ from -55 to $+200°C$)
4. $\pm$ 8% tolerance on permeability
5. Very small change in permeability with variation in excitation level (less than 5% change in permeability from 0.001 to 0.2 T)
6. Curie temperature of 450°C
7. Low core losses (less than 20 W/kg at 50 kHz, 0.1 T)
8. MPP cores withstand large direct currents without saturating the core (14-μ cores have a 20% reduction in μ at 200 Oe of dc current)

---

*Sections 5.1.1 and 5.1.2 have been contributed by Bob Noah of Magnetics Inc. and Jim Cox of Micrometals, respectively.

Molypermalloy power cores are widely used in high-reliability military and space applications because of their good stability over wide temperature ranges and their ability to withstand high levels of shock, vibration, and nuclear radiation without degradation. Other applications for these cores are:

1. Stable, high-Q filters operating in the frequency range 1 kHz to 1 MHz
2. Loading coils used to cancel out the distributed capacitance in telephone cables
3. Pulse transformers
4. Differential mode EMI noise filters
5. Flyback transformers
6. Energy storage or output chokes in circuits with large amounts of dc current flowing

The balance of this chapter deals extensively with core selection and the design of chokes carrying dc current with ac ripple present.

5.1.2 Iron Powder Cores

In the last century the use of solid magnetic material for dc electromagnetics and later laminated magnetic materials for low-frequency applications led to the need for materials that would operate efficiently at higher and higher frequencies. With the original thick laminations it was discovered that the apparent permeability or inductance decreased as frequency increased, and at the same time losses became prohibitive. It was found that when thin sheets of material insulated from one another were used, better results were obtained. This is primarily due to an effect known as *eddy current shielding*. As frequency increases, the depth of magnetic penetration decreases for any given material. Thus by having thin sheets, more of the core body is utilized. Thinner laminations and grain-oriented alloys meet the higher frequency needs.

While the thin oriented laminations were useful for broadband audio transformers, they were unsuitable for selective circuits where high Q is required. While at low frequency the magnetic field in a coil is in its axial direction, at high frequency each turn generates its own field concentric with the wire. These fields are coupled with fields from adjacent turns and are coupled to the core through axial fields rather than one central field. This type of field requires cores laminated in all directions in order to minimize losses and thus maintain reliable inductance and Q versus frequency. As a result, powdered iron cores were developed.

There are two basic classes of iron powders available: hydrogen-reduced irons and carbonyl irons. The hydrogen-reduced iron powder has low resistance and a relatively large particle size. It produces the highest permeabilities (approaching 100) and has low losses at low frequency, but the losses increase significantly at high frequency, producing very low Q at radio frequencies. Cores from this powder are output chokes in switching power supplies.

The carbonyl irons, on the other hand, have particles that are formed by the decomposition of pentacarbonyliron vapor, which produces a spherical particle with an onion-skin structure. The laminating effect of the "onion skin" produces a resistivity of the individual particles that is much higher than that of pure iron. This high resistance in conjunction with the very small particle size (3–5 µm) greatly enhances high-frequency performance. The permeability of carbonyliron powders, and thus their inductance, can be manufactured to a very tight tolerance and the powders remain extremely stable with frequency, temperature, and applied signal level. All of these are important considerations in high-Q selective circuits.

The distributed air gap characteristic of iron powder produces a core with permeabilities ranging from 4 to 100. This feature in conjunction with the inherent high saturation point of iron makes it very difficult to saturate.

While iron powder cores may be limited in their use because of low permeabilities or rather high core losses at higher frequencies, they have become a very popular choice as a core material for high-volume commercial applications due to their low cost compared with other core materials.

5.2 INDUCTORS

Inductors that carry direct current are used frequently in a wide variety of ground, air, and space applications. Selection of the best magnetic core for an inductor frequently involves a trial-and-error type of calculation.

The design of an inductor also frequently involves consideration of the effect of its magnetic field on other devices in its immediate vicinity. This is especially true in the design of high-current inductors for converters and switching regulators used in spacecraft, which may also employ sensitive magnetic field detectors. For this type of design problem it is frequently imperative that a toroidal core be used. The magnetic flux in a Molypermalloy toroid (core) can be contained inside the core more readily than in a lamination or C core, as the winding covers the core along the entire magnetic path length.

The author has developed a simplified method of designing optimum dc carrying inductors with Molypermalloy powder cores. This method allows the correct core permeability to be determined without relying on trial and error.

5.2.1 Relationship of A_p to Inductor's Energy-Handling Capability

According to the newly developed approach, the energy-handling capability of a core is related to its area product A_p by the equation

$$A_p = \left(\frac{2(\text{Energy}) \times 10^4}{B_m K_u K_j} \right)^{1.14} \quad [\text{cm}^4] \tag{5.1}$$

where K_j is the current density coefficient (see Chapter 2)

K_u is the window utilization factor (see Chapter 6)

B_m is the flux density, tesla

Energy is in watt seconds

From the above, it can be seen that factors such as flux density, window utilization factor K_u (which defines the maximum space that may be occupied by the copper in the window), and the constant K_j (which is related to temperature rise) all have an influence on the inductor area product. The constant K_j is a new parameter that gives the designer control of the copper losses. Derivation is set forth in detail in Chapter 2. The energy-handling capability of a core is derived from

$$\text{Energy} = \frac{LI^2}{2} \quad [\text{W-s}] \tag{5.2}$$

5.2.2 Relationship of K_g to Inductor's Energy-Handling Capability

Inductors, like transformers, are designed for a given temperature rise. They can also be designed for a given regulation. The regulation and energy handling ability of a core is related to two constants:

$$*\alpha = \frac{(\text{Energy})^2}{K_g K_e} \quad [\%] \tag{5.3}$$

where α is the regulation, %.

The constant K_g is determined by the core geometry:

$$*K_g = \frac{W_a A_c^2 K_u}{\text{MLT}} \quad [\text{cm}^5] \tag{5.4}$$

The constant K_e is determined by the magnetic and electrical operating conditions:'

$$*K_e = 0.145 P_o B_{\max}^2 \times 10^{-4} \tag{5.5}$$

where P_o is output power and

$$B_{\max} = B_{\text{dc}} + \frac{B_{\text{ac}}}{2} \tag{5.6}$$

From the above it can be seen that flux density is the predominant factor governing size.

5.3 FUNDAMENTAL CONSIDERATIONS

The design of a linear reactor depends upon four related factors:

1. Desired inductance
2. Direct current

---

*The derivation for these equations are at the end of Chapter 4.

Table 5.1. Powder core permeabilities

Permeability	Amp turn/cm with dc bias $L < 80\%$
14	253
26	140
60	56
125	28
147	23
160	20
173	19
200	16
300	11
555	4

3. Alternating current ΔI
4. Power loss and temperature rise

With these requirements established, the designer must determine the maximum values for B_{dc} and for B_{ac} that will not produce magnetic saturation, and must make tradeoffs that will yield the highest inductance for a given volume. The core permeability chosen dictates the maximum dc flux density that can be tolerated for a given design. Permeability values for a number of powder cores are shown in Table 5.1.

If an inductance is to be constant with increasing direct current, there must be a negligible drop in inductance over the operating current range. The maximum H, then, in an indication of a core's capability. In terms of ampere-turns and mean magnetic path length l_m,

$$H = \frac{NI}{l_m} \quad [\text{amp-turn/cm}] \tag{5.7}$$

$$NI = 0.8Hl_m \quad [\text{amp-turn}] \tag{5.8}$$

Inductance decreases with increasing flux density and magnetizing force for various materials of different values of permeability μ_Δ. The selection of the correct permeability for a given design is made using Equation (5.8) after solving for the area product A_p:*

$$\mu_\Delta = \frac{B_m l_m \times 10^4}{0.4\pi W_a J K_u} \tag{5.9}$$

It should be remembered that maximum flux density depends upon $B_{dc} + B_{ac}$ in the manner shown in Figure 5.1.

---

*Derivation is set forth in detail in Appendix A at the end of this chapter.

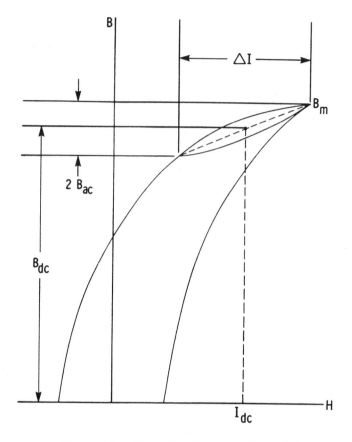

Figure 5.1. Flux density versus $I_{dc} + \Delta I$.

$$B_m = B_{dc} + B_{ac} \quad [\text{T}] \tag{5.10}$$

$$B_{dc} = \frac{0.4\pi NI_{dc} \times 10^{-4}}{l_g + l_m/\mu_r} \quad [\text{T}] \tag{5.11}$$

$$B_{ac} = \frac{0.4\pi N(\Delta I/2) \times 10^{-4}}{l_g + l_m/\mu_r} \quad [\text{T}] \tag{5.12}$$

Combining Equations (5.11) and (5.12),

$$B_m = \frac{0.4\pi NI_{dc} \times 10^{-4}}{l_g + l_m/\mu_r} + \frac{0.4\pi N(\Delta I/2) \times 10^{-4}}{l_g + l_m/\mu_r} \quad [\text{T}] \tag{5.13}$$

Molypermalloy powder cores operating with a dc bias of 0.3 T have only about 80% of their original inductance, with very rapid falloff at higher densities as shown in Figure 5.2.

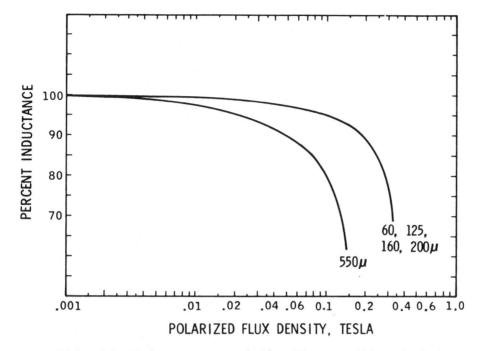

Figure 5.2. Inductance versus dc bias. (*Courtesy of Magnetics Inc.*)

The flux density for the initial design for Molypermalloy powder cores should be limited to 0.3 T maximum for B_{dc} plus B_{ac}.

The losses in a Molypermalloy inductor due to ac flux density are very low compared to the steady-state dc copper loss. It is then assumed that the majority of the losses are copper:

$$P_{cu} \gg P_{fe} \qquad (5.14)$$

5.4 TABLES FOR MPP MOLYPERMALLOY POWDER CORES AND POWDERED IRON CORES

The author has developed Tables 5.2 and 5.3 to be used as a tool to simplify and standardize the process of inductor design. This will make it possible to design inductors of lighter weight and smaller volume and to optimize efficiency without going through a cut-and-try design procedure. While developed specifically for aerospace applications, the information has wider utility and can be used for the design of non-aerospace inductors as well. A great deal of information is presented for the convenience of the designer. The material is in tabular form to assist the designer in making the tradeoffs best suited for a particular application in a minimum amount of time.

The core data for these tables were taken from Ref. 1. The cores chosen were just a small sample of over 300 cores available. In Table 5.2 are 16 standard 60 perm MPP Molypermalloy powder cores and in Table 5.3 are 16 standard 75 perm powdered iron cores that are most frequently used in switch mode power supplies.

Table 5.2. Micrometals Powered Iron Cores

Cat. no.	Bare i.d. 1	Wound o.d. 2	Wound HT 3	MPL 4	WTFE 5	WTCU 6	MLT 7
T30-26	0.384	0.975	0.517	1.83	0.800	0.47	1.14
T37-26	0.521	1.213	0.585	2.32	1.100	0.97	1.28
T44-26	0.582	1.412	0.695	2.67	2.000	1.46	1.54
T50-26	0.770	1.650	0.868	3.20	2.700	2.96	1.79
T68-26	0.940	2.220	0.953	4.24	5.700	5.36	2.17
T72-26	0.711	2.220	1.015	3.99	10.400	3.56	2.52
T80-26	1.260	2.640	1.265	5.15	8.900	11.67	2.63
T94-26	1.420	3.091	1.452	6.00	16.100	17.45	3.10
T106-26	1.450	3.414	1.835	6.50	32.400	23.07	3.93
T130-26	1.980	4.277	2.100	8.29	43.200	48.35	4.42
T150-26	2.150	4.908	2.185	9.41	62.100	62.59	4.85
T157-26	2.410	5.178	2.655	10.05	80.200	89.41	5.51
T200-26	3.180	6.645	2.990	12.97	122.800	178.04	6.30
T184-26	2.410	5.882	3.005	11.12	155.500	107.32	6.62
T201-26	2.410	6.316	3.405	11.76	245.500	123.02	7.58
T300-26	4.890	10.146	3.715	19.83	254.800	549.23	8.22

Table 5.3. Magnetics Inc. MPP Powder Cores

Cat. no.	Bare i.d. 1	Wound o.d. 2	Wound HT 3	MPL 4	WTFE 5	WTCU 6	MLT 7
55291	.376	1.297	.696	2.18	1.800	.66	1.68
55041	.406	1.362	.711	2.38	1.900	.79	1.72
55051	.648	1.731	.926	3.12	3.100	2.44	2.08
55121	.902	2.247	1.213	4.11	6.800	6.03	2.65
55848	1.156	2.739	1.340	5.09	10.000	11.00	2.95
55059	1.288	3.055	1.533	5.67	16.000	15.53	3.35
55351	1.326	3.144	1.684	5.88	20.000	17.77	3.62
55894	1.359	3.515	1.929	6.35	36.000	21.96	4.26
55586	2.209	4.659	2.138	8.95	35.000	61.48	4.51
55071	1.879	4.367	2.151	8.15	47.000	46.19	4.68
55076	2.099	4.762	2.229	8.98	52.000	59.84	4.86
55083	2.279	5.255	2.727	9.84	92.000	84.68	5.84
55090	2.739	6.169	3.033	11.63	131.000	136.43	6.51
55716	3.039	6.724	3.005	12.73	133.000	169.06	6.55
55439	2.279	5.980	3.083	10.74	182.000	100.93	6.96
55110	3.419	7.541	3.246	14.30	176.000	233.10	7.14

A_c 2 8	W_a 2 9	A_p 4 10	K_g 5 11	A_t 2 12	Perm (U) 13	MH/1000 turns 14
0.065	0.116	0.008	0.000171	2.5	75.0	33.0
0.070	0.213	0.015	0.000326	3.7	75.0	28.0
0.107	0.266	0.028	0.000790	5.1	75.0	36.0
0.121	0.466	0.056	0.001525	7.0	75.0	32.0
0.196	0.694	0.136	0.004908	11.5	75.0	42.0
0.369	0.397	0.147	0.008581	12.3	75.0	87.0
0.242	1.247	0.302	0.011098	16.9	75.0	45.0
0.385	1.584	0.610	0.030297	23.1	75.0	59.0
0.690	1.651	1.139	0.080059	31.0	75.0	90.0
0.733	3.079	2.257	0.149851	45.3	75.0	79.0
0.937	3.631	3.402	0.262993	56.9	75.0	94.0
1.140	4.562	5.200	0.430214	68.2	75.0	97.0
1.330	7.942	10.563	0.891438	103.0	75.0	90.0
2.040	4.562	9.306	1.147754	89.8	75.0	164.0
2.940	4.562	13.411	2.079602	108.1	75.0	226.0
1.810	18.781	33.993	2.992552	212.5	75.0	83.0

A_c 2 8	W_a 2 9	A_p 4 10	K_g 5 11	A_t 2 12	Perm (U) 13	MH/1000 turns 14
.094	.111	.010	.000237	4.7	60.0	32.0
.100	.129	.013	.000301	5.1	60.0	32.0
.114	.330	.038	.000824	8.1	60.0	27.0
.192	.639	.123	.003553	13.6	60.0	35.0
.226	1.050	.237	.007274	18.9	60.0	32.0
.331	1.303	.431	.017039	23.8	60.0	43.0
.388	1.381	.536	.022982	26.3	60.0	51.0
.654	1.451	.949	.058299	33.8	60.0	75.0
.454	3.832	1.740	.070042	51.3	60.0	38.0
.672	2.773	1.863	.106936	48.0	60.0	61.0
.678	3.460	2.346	.130831	55.1	60.0	56.0
1.072	4.079	4.373	.321214	71.7	60.0	81.0
1.340	5.892	7.895	.649954	95.1	60.0	86.0
1.251	7.254	9.074	.692776	106.3	60.0	73.0
1.990	4.079	8.118	.928721	94.3	60.0	135.0
1.444	9.181	13.257	1.072468	130.7	60.0	75.0

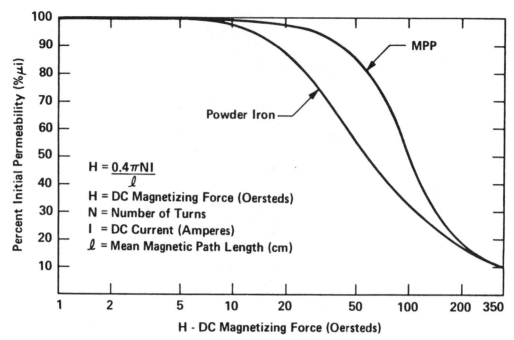

Figure 5.3.   Percent initial permeability vs. dc magnetizing force.

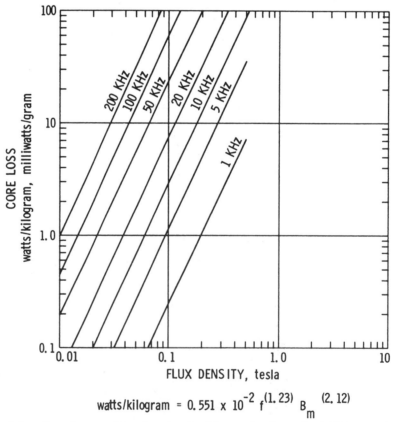

$$\text{watts/kilogram} = 0.551 \times 10^{-2}\, f^{(1.23)}\, B_m^{(2.12)}$$

Figure 5.4.   Core loss vs flux density for Magnetics Inc. 60μ MPP powder core.

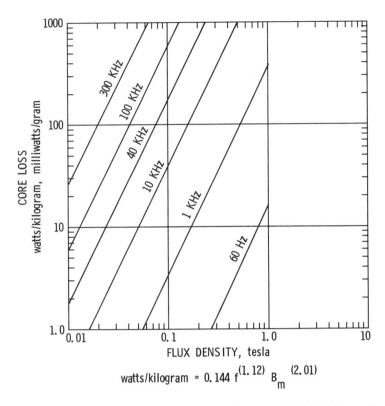

Figure 5.5. Core loss vs flux density for Micrometals 75μ 26 powder iron.

Figure 5.3 is a comparison of 60 perm Molypermalloy MPP and 75 perm powder iron permeability versus dc bias. The core loss data for 60 perm MPP Molypermalloy powder cores and 75 perm powdered iron cores are shown in Figures 5.4 and 5.5. Care must be taken in designing inductors with a dc bias because of the loss of permeability due to the high dc magnetizing force. The design work sheets on the following pages will provide a direct approach to the design of MPP powder cores using either the area product A_p or the core geometry K_g. These work sheets will function just as well for designing inductors with powdered iron cores. Powder iron cores are well suited for input filter inductors in switch mode power supplies.

5.5 TOROIDAL INDUCTOR DESIGN USING AREA PRODUCT A_p APPROACH

Design with the following specification:

1. Inductance, $L = 0.003$ [H]
2. dc current, $I_o = 3$ [A]
3. ac current, $\Delta I = 0.2$ [A]
4. Ripple frequency, $f = 20,000$ [Hz]

5. Flux density, $B_m = 0.3$ [T]
6. Temperature rise = 25   [°C]
7. Core material: MPP
8. Core configuration: toroidal powder core

Step 1.   Calculate the energy-handling capability:

$$\text{Energy} = \frac{LI^2}{2} \quad \text{[W-s]}$$

$$I = I_o + \frac{\Delta I}{2} \quad \text{[A]}$$

$$I = 3 + \frac{0.2}{2}$$

$$I = 3.1 \quad \text{[A]}$$

$$\text{Energy} = \frac{(0.003)(3.1)^2}{2}$$

$$\text{Energy} = 0.0144 \quad \text{[W-s]}$$

Step 2.   Calculate the area product A_p:

$$A_p = \left(\frac{2(\text{Energy}) \times 10^4}{B_m K_u K_j} \right)^{(x)} \quad \text{[cm}^4\text{]}$$

Window utilization factor $K_u = 0.4$

$$A_p = \left(\frac{2(0.0144) \times 10^4}{(0.3)(0.4)(403)} \right)^{1.14}$$

$$A_p = 7.6 \quad \text{[cm}^4\text{]}$$

Step 3.   Select a comparable area product A_p for the specified core section and record the appropriate data.

Core 55441

$A_p = 8.2$ [cm^4]

$\text{MLT} = 6.96$ [cm]

$A_c = 1.99$ [cm^2]

$W_a = 4.08$ [cm^2]

$A_t = 94.2$ [cm^2]

$\text{MPL} = 10.74$ [cm]

$W_{tfe} = 182$ [g]

Step 4.   Calculate the current density J using K_j and y from Table 3.1:

$$J = K_j A_p^{(y)}\quad [\text{A/cm}^2]$$

$$J = (403)(8.2)^{-0.12}$$

$$J = 313\quad [\text{A/cm}^2]$$

Step 5.   Calculate the bare wire size $A_{w(B)}$:

$$A_{w(B)} = \frac{I_o + \Delta I/2}{J}\quad [\text{cm}^2]$$

$$A_{w(B)} = \frac{3 + 0.2/2}{313}$$

$$A_{w(B)} = 0.0102 \quad [\text{cm}^2]$$

Step 6. Select a wire size from (Table 6.1), column 2. If the area is not within 10%, take the next smallest size. Also record micro-ohms per centimeter from column 4.

AWG No. 17

$$\text{Bare}, A_{w(B)} = 0.01039 \quad [\text{cm}^2]$$

$$\text{Insulated}, A_w = 0.01168 \quad [\text{cm}^2]$$

$$\mu\Omega/\text{cm} = 166$$

Step 7. Calculate the effective window area $W_{a(\text{eff})}$. Use window area W_a found in step 3:

$$W_{a(\text{eff})} = W_a S_3 \quad [\text{cm}^2]$$

A typical value for S_3 is 0.75, as shown in Chapter 6.

$$W_{a(\text{eff})} = (4.08)(0.75)$$

$$W_{a(\text{eff})} = 3.06 \quad [\text{cm}^2]$$

Step 8. Calculate N, the number of turns. Use wire area A_w found in step 6:

$$N = \frac{W_{a(\text{eff})} S_2}{A_w} \quad [\text{turns}]$$

A typical value for S_2 is 0.6, as shown in Chapter 6.

$$N = \frac{(3.06)(0.6)}{0.0117}$$

$$N = 157 \quad [\text{turns}]$$

Step 9. Calculate the permeability of the core required:

$$\mu_r = \frac{L \times \text{MPL} \times 10^8}{0.4\pi N^2 A_c}$$

$$\mu_r = \frac{(0.003)(10.74) \times 10^8}{(1.26)(157)^2(1.99)}$$

$$\mu_r = 52$$

Now choose a core from the group with a closer permeability.

No. 55439-A2

$$\mu = 60$$

$$\text{MH}/1000 = 135$$

Step 10. Calculate the number of turns required:

$$N = 1000\left(\frac{L}{L_{1000}}\right) \quad [\text{turns}]$$

$$N = 1000\left(\frac{3}{135}\right)$$

$$N = 149 \quad [\text{turns}]$$

Step 11. Calculate the winding resistance. Use MLT from step 3 and micro-ohms per centimeter from step 6.

$$R = (\text{MLT})(N)\left(\frac{\mu\Omega}{\text{cm}}\right) \times 10^{-6} \quad [\Omega]$$

$$R = (6.96)(149)(166) \times 10^{-6}$$

$$R = 0.157 \quad [\Omega]$$

Step 12.   Calculate the copper loss P_{cu}:

$$P_{cu} = I^2R \quad [W]$$

$$P_{cu} = (3.1)^2(0.157)$$

$$P_{cu} = 1.51 \quad [W]$$

Step 13.   Calculate the ac flux density:

$$B_{ac} = \frac{0.4\pi N(\Delta I/2)\mu \times 10^{-4}}{MPL} \quad [T]$$

$$B_{ac} = \frac{(1.26)(149)(0.1)(60) \times 10^{-4}}{10.74}$$

$$B_{ac} = 0.0105 \quad [T]$$

Step 14.   Calculate the watts per kilogram for the appropriate core material; then determine the core loss. Core weight is found in step 3:

$$W/kg = Kf^{(m)}B_m^{(n)} \quad [T]$$

$$mW/g = (0.551)(20,000)^{1.23}(0.0105)^{2.12} \times 10^{-2}$$

$$mW/g = 0.0686$$

$$P_{fe} = (mW/g)(W_{tfe}) \times 10^{-3} \quad [W]$$

$$P_{fe} = (0.0686)(182) \times 10^{-3}$$

$$P_{fe} = 0.0125 \quad [W]$$

Step 15. Calculate the total losses P_Σ:

$$P_\Sigma = P_{cu} + P_{fe} \quad [W]$$

$$P_\Sigma = 1.51 + 0.0125$$

$$P_\Sigma = 1.52 \quad [W]$$

Step 16. Calculate the watts per unit area. The surface area A_t is found in step 4:

$$\psi = \frac{P_{cu}}{A_t} \quad [W/cm^2]$$

$$\psi = \frac{1.52}{94.2}$$

$$\psi = 0.0161 \quad [W/cm^2]$$

Step 17. Calculate the dc magnetizing force (oersteds):

$$H = \frac{0.4\pi NI}{MPL} \quad [Oe]$$

$$H = \frac{(1.26)(149)(3.1)}{10.74}$$

$$H = 54.2 \quad [Oe]$$

5.6 TOROIDAL INDUCTOR DESIGN USING THE CORE GEOMETRY K_g APPROACH

Design with the following specification:

1. Inductance, $L = 0.0025 \quad [H]$

2. dc current, $I_o = 1.5$ [A]
3. ac current, $\Delta I = 0.2$ [A]
4. Output power, $P_o = 100$ [W]
5. Regulation, $\alpha = 1.0$ [%]
6. Ripple frequency, $f = 20,000$ [Hz]
7. Flux density, $B_m = 0.3$ [T]
8. Core material: MPP
9. Core configuration: toroid

Step 1.   Calculate the energy-handling capability:

$$\text{Energy} = \frac{LI^2}{2} \quad [\text{W-s}]$$

$$I = I_o + \frac{\Delta I}{2} \quad [\text{A}]$$

$$I = 1.5 + \frac{0.2}{2}$$

$$I = 1.6 \quad [\text{A}]$$

$$\text{Energy} = \frac{(0.0025)(1.6)^2}{2}$$

$$\text{Energy} = 0.0032 \quad [\text{W-s}]$$

Step 2.   Calculate the electrical conditions constant K_e:

$$K_e = 0.145 P_o B_m^2 \times 10^{-4}$$

$$K_e = (0.145)(100)(0.3)^2 \times 10^{-4}$$

$$K_e = 0.0013$$

Step 3.   Calculate the core geometry coefficient K_g:

$$K_g = \frac{(\text{Energy})^2}{K_e \alpha} \quad [\text{cm}^5]$$

$$K_g = \frac{(0.0032)^2}{(0.00013)(1)}$$

$$K_g = 0.0788 \quad [\text{cm}^5]$$

Step 4. Select a comparable core geometry K_g for the specified core, and record the appropriate data.

Core 55583

$$K_g = 0.07 \quad [\text{cm}^5]$$

$$A_p = 1.74 \quad [\text{cm}^4]$$

$$\text{MLT} = 4.51 \quad [\text{cm}]$$

$$A_c = 0.454 \quad [\text{cm}^2]$$

$$W_a = 3.83 \quad [\text{cm}^2]$$

$$A_t = 51.3 \quad [\text{cm}^2]$$

$$\text{MPL} = 8.95 \quad [\text{cm}]$$

$$W_{tfe} = 35 \quad [\text{g}]$$

Step 5. Calculate the current density J. Use area product A_p found in step 4:

$$J = \frac{2(\text{Energy}) \times 10^4}{B_m A_p K_u} \quad [\text{A/cm}^2]$$

Window utilization factor, $K_u = 0.4$

$$J = \frac{(2)(0.0032) \times 10^4}{(0.3)(1.74)(0.4)}$$

$$J = 306 \quad [\text{A/cm}^2]$$

Step 6.   Calculate the bare wire size $A_{w(B)}$:

$$A_{w(B)} = \frac{I_o + \Delta I/2}{J} \quad [\text{cm}^2]$$

$$A_{w(B)} = \frac{1.5 + 0.2/2}{306}$$

$$A_{w(B)} = 0.00523 \quad [\text{cm}^2]$$

Step 7.   Select a wire size from Table 6.1, column 2. If the area is not within 10%, take the next smallest size. Also record micro-ohms per centimeter from column 4.

AWG No. 20

Bare, $A_{w(B)} = 0.00519 \quad [\text{cm}^2]$

Insulated, $A_w = 0.00606 \quad [\text{cm}^2]$

$\mu\Omega/\text{cm} = 332$

Step 8.   Calculate the effective window area $W_{a(\text{eff})}$. Use window area W_a found in step 4:

$$W_{a(\text{eff})} = W_a S_3 \quad [\text{cm}^2]$$

A typical value for S_3 is 0.75, as shown in Chapter 6.

$$W_{a(\text{eff})} = (3.83)(0.75)$$

$$W_{a(\text{eff})} = 2.87 \quad [\text{cm}^2]$$

Step 9.   Calculate N, the number of turns. Use wire area A_w found in step 7:

$$N = \frac{W_{a(\text{eff})}S_2}{A_w} \quad [\text{turns}]$$

A typical value for S_2 is 0.6, as shown in Chapter 6.

$$N = \frac{(2.87)(0.6)}{0.00606}$$

$$N = 284 \quad [\text{turns}]$$

Step 10.   Calculate the permeability of the core required:

$$\mu_r = \frac{L \times \text{MPL} \times 10^8}{0.4\pi N^2 A_c}$$

$$\mu_r = \frac{(0.0025)(8.95) \times 10^8}{(1.26)(284)^2(0.454)}$$

$$\mu_r = 48$$

Now choose a core from the group with a closer permeability.

No. 55586-A2

$$\mu = 60$$

$$\text{MH}/1000 = 38$$

Step 11. Calculate the number of turns required:

$$N = 1000\left(\frac{L}{L_{1000}}\right) \quad \text{[turns]}$$

$$N = 1000\left(\frac{2.5}{38}\right)$$

$$N = 256 \quad \text{[turns]}$$

Step 12. Calculate the winding resistance. Use MLT from step 4 and micro-ohms per centimeter from step 7.

$$R = (\text{MLT})(N)\left(\frac{\mu\Omega}{\text{cm}}\right) \times 10^{-6} \quad [\Omega]$$

$$R = (451)(256)(332) \times 10^{-6}$$

$$R = 0.383 \quad [\Omega]$$

Step 13. Calculate the copper loss P_{cu}:

$$P_{cu} = I^2 R \quad \text{[W]}$$

$$P_{cu} = (1.6)^2(0.383)$$

$$P_{cu} = 0.980 \quad \text{[W]}$$

Step 14. Calculate the regulation, α:

$$\alpha = \frac{P_{cu}}{P_o} \times 100 \quad \text{[\%]}$$

$$\alpha = \frac{0.98 \times 100}{100}$$

$$\alpha = 0.98 \quad [\%]$$

Step 15. Calculate the ac flux density:

$$B_{ac} = \frac{0.4\pi N(\Delta I/2)\mu \times 10^{-4}}{MPL} \quad [T]$$

$$B_{ac} = \frac{(1.26)(256)(0.1)(60) \times 10^{-4}}{8.95}$$

$$B_{ac} = 0.0216 \quad [T]$$

Step 16. Calculate the watts per kilogram for the appropriate core material; then determine the core loss. Core weight is found in step 4:

$$W/kg = Kf^{(m)}B_m^{(m)}$$

$$mW/g = (0.551)(20,000)^{1.23}(0.0216)^{2.12} \times 10^{-2}$$

$$mW/g = 0.317$$

$$P_{fe} = (mW/g)(W_{tfe}) \times 10^{-3} \quad [W]$$

$$P_{fe} = (0.316)(35) \times 10^{-3}$$

$$P_{fe} = 0.0111 \quad [W]$$

Step 17. Calculate the total losses P_Σ:

$$P_\Sigma = P_{cu} + P_{fe} \quad [W]$$

$$P_\Sigma = 0.98 + 0.0111$$

$$P_\Sigma = 0.99 \quad [W]$$

Step 18.   Calculate the watts per unit area. The surface area A_t is found in step 4:

$$\psi = \frac{P_{cu}}{A_t} \quad [\text{W/cm}^2]$$

$$\psi = \frac{0.99}{51.3}$$

$$\psi = 0.0193 \quad [\text{W/cm}^2]$$

Step 19.   Calculate the dc magnetizing force (oersteds):

$$H = \frac{0.4\pi NI}{\text{MPL}} \quad [\text{Oe}]$$

$$H = \frac{(1.26)(256)(1.6)}{8.95}$$

$$H = 57.6 \quad [\text{Oe}]$$

REFERENCE

1.   McLyman, C. W. T., *Magnetic Core Selection for Transformers and Inductors,* Marcel Dekker, New York, 1982.

BIBLIOGRAPHY

Blinchikoff, H., Toroidal Inductor Design, Electro-Technology, November 1964, pp. 42–50.

Smith, G. D., Designing Toroidal Inductors with dc Bias. NASA Technical Note D-2320, Goddard Space Flight Center, Greenbelt, Maryland.

Stan, P., Toroid Design Analysis. Electro-Technology, August 1966, pp. 85–94.

APPENDIX A. TOROIDAL POWDER CORE SELECTION WITH DC CURRENT

After calculating the inductance and dc current, select the proper permeability and size of powder core with a given $LI^2/2$. The enerlgy-handling capability of an inductor can be determined by its A_p product. The relationship is derived by solving $E = L\, dI/dt$ as follows:*

$$E = L\frac{dI}{dt} = N\frac{d\phi}{dt} \tag{5.A-1}$$

$$L = N\frac{d\phi}{dI} \tag{5.A-2}$$

$$\phi = B_m A_c' \tag{5.A-3}$$

$$B_m = \mu_\Delta \mu_0 H = \frac{\mu_\Delta \mu_0 NI}{l_m'} \tag{5.A-4}$$

$$\phi = \frac{\mu_\Delta \mu_0 NIA_c'}{l_m'} \tag{5.A-5}$$

$$\frac{d\phi}{dI} = \frac{\mu_\Delta \mu_0 NA_c'}{l_m'} \tag{5.A-6}$$

$$L = N\frac{d\phi}{dI} = \frac{\mu_\Delta \mu_0 N^2 A_c'}{l_m'} \tag{5.A-7}$$

$$\text{Energy} = \frac{LI^2}{2} = \frac{\mu_\Delta \mu_0 N^2 A_c' I^2}{2l_m'} \tag{5.A-8}$$

If B_m is specified,

$$I = \frac{B_m l_m'}{\mu_\Delta \mu_0 N} \tag{5.A-9}$$

---

*Primes indicate measurements in the mks system.

$$\text{Energy} = \frac{\mu_\Delta \mu_0 N^2 A_c'}{2 l_m'} \left(\frac{B_m l_m'}{\mu_\Delta \mu_0 N} \right)^2 \qquad (5.\text{A-}10)$$

Reducing to

$$\text{Energy} = \frac{B_m^2 l_m' A_c'}{2 \mu_\Delta \mu_0} \quad [\text{W-s}] \qquad (5.\text{A-}11)$$

$$I = \frac{K_u W_a' J'}{N} = \frac{B_m l_m'}{\mu_\Delta \mu_0 N} \qquad (5.\text{A-}12)$$

Solving for $\mu_\Delta \mu_0$:

$$\mu_\Delta \mu_0 = \frac{\dot{B}_m l_m'}{K_u W_a' J'} \qquad (5.\text{A-}13)$$

Substituting into the energy equation,

$$\text{Energy} = \frac{B_m^2 l_m' A_c'}{2} \left(\frac{K_u W_a' J'}{B_m l_m'} \right) = \frac{W_a' A_c' B_m J' K_u}{2} \qquad (5.\text{A-}14)$$

Let $l_m' = l_m \times 10^{-2}$
$\quad\ W_a' = W_a \times 10^{-4}$
$\quad\ A_c' = A_c \times 10^{-4}$
$\quad\ J' = J \times 10^4$

Substituting into the energy equation,

$$\text{Energy} = \frac{W_a A_c B_m J K_u}{2} \times 10^{-4} \qquad (5.\text{A-}15)$$

Solving for $W_a A_c$,

$$W_a A_c = \frac{2(\text{Energy}) \times 10^4}{K_u B_m J} \qquad (5.\text{A-}16)$$

Since the area product is

$$A_p = W_a A_c \tag{5.A-17}$$

we have

$$A_p = \frac{2(\text{Energy}) \times 10^4}{K_u B_m J} \tag{5.A-18}$$

Combining the equation from Table 2.1,

$$J = K_j A_p^{-0.12} \tag{5.A-19}$$

yielding

$$A_p = \frac{2(\text{Energy}) \times 10^4}{K_u B_m (K_j A_p^{-0.12})} \tag{5.A-20}$$

$$A_p^{0.88} = \frac{2(\text{Energy}) \times 10^4}{K_u B_m K_j} \tag{5.A-21}$$

$$A_p = \left(\frac{2(\text{Energy}) \times 10^4}{K_u B_m K_j} \right)^{1.14} \quad [\text{cm}^4] \tag{5.A-22}$$

After the core size has been determined, the next step is to pick the right permeability for that core size. This is done by solving for μ_Δ in Equation (5.A-13).

$$\mu_\Delta = \frac{B_m l_m \times 10^{-2}}{\mu_0 W_a J K_u} \tag{5.A-23}$$

for $\mu_0 = 4\pi \times 10^{-7}$,

$$\mu_\Delta = \frac{B_m l_m \times 10^4}{0.4\pi W_a J K_u} \tag{5.A-24}$$

APPENDIX B. MAGNETIC AND DIMENSIONAL SPECIFICATIONS FOR 13 COMMONLY USED MOLYPERMALLOY CORES

The following remarks apply to each of Tables 5.4–5.16, which were compiled from manufacturers' data.

1. Total weight − core weight plus wire weight, assuming AWG 20
2. Maximum OD of wound core with residual hole = 1/2 ID

Table 5.4. Dimensional specifications for Magnetics Inc. 55051-A2 and Arnold Engineering A-051027-2

	ENGLISH	METRIC
Wa/Ac		3.39
Wa x Ac	0.00104 in^4	0.0432 cm^4
OD	0.530 in	1.346 cm
ID	0.275 in	0.699 cm
HT	0.217 in	0.551 cm
Wa = WINDOW AREA	0.075 x 10^6 CIR-MIL	0.383 cm^2
Wa = EFFECTIVE	0.0445 in^2	0.288 cm^2
Ac = CROSS SECTION	0.0175 in^2	0.113 cm^2
lm = PATH LENGTH	1.229 in	3.12 cm
CORE WEIGHT	0.0066 lb	3.0 grams
TOTAL WEIGHT	0.0106 lb	5.175 grams
WOUND OD MIN	0.581 in	1.475 cm
MLT	0.850 in	2.160 cm
A$_t$ = SURFACE AREA	1.018 in^2	6.568 cm^2
PERMEABILITY		60
μ 125		2.08 x L @ μ 60
μ 160		2.67 x L @ μ 60
μ 200		3.33 x L @ μ 60
μ 550		9.17 x L @ μ 60

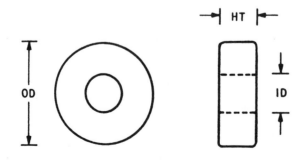

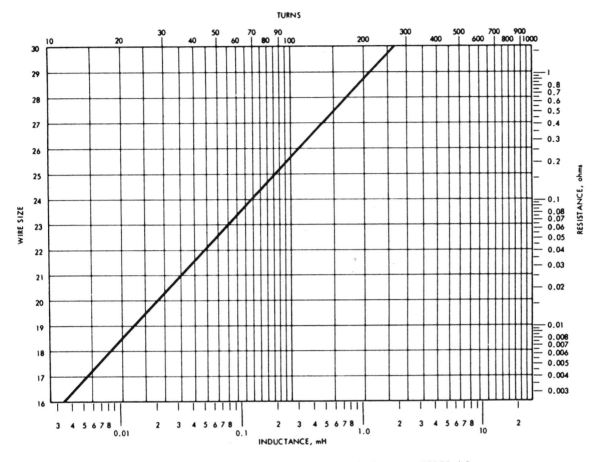

Figure 5.6. Wire and inductance graph for core 55051-A2.

3. MLT (mean length/turn) full-wound toroid
4. Effective window area $W_{(eff)} = 3\pi r^2/4$

The graphs in Figure 5.6–5.18 relate to the 13 different core sizes. They show resistance, number of turns, inductance, and wire size for a window utilization factor of 0.40 and are based on a permeability of 60. To convert for other permeability values, the appropriate inductance multiplication factors listed should be used. The information appearing in the tables and figures will enable the engineer to arrive at a close approximation for breadboarding purposes.

Table 5.5. Dimensional specifications for Magnetics Inc. 55121-A2 and Arnold Engineering A-266036-2

	ENGLISH	METRIC
Wa/Ac		3.63
Wa x Ac	0.00336 in^4	0.139 cm^4
OD	0.680 in	1.740 cm
ID	0.375 in	0.953 cm
HT	0.280 in	0.711 cm
Wa = WINDOW AREA	0.141 x 10^6 CIR-MIL	0.713 cm^2
Wa = EFFECTIVE	0.0828 in^2	0.535 cm^2
Ac = CROSS SECTION	0.0304 in^2	0.196 cm^2
lm = PATH LENGTH	1.62 in	4.11 cm
CORE WEIGHT	0.0143 lb	6.50 grams
TOTAL WEIGHT	0.0257 lb	11.70 grams
WOUND OD MIN	0.753 in	1.925 cm
MLT	1.075 in	2.74 cm
A$_t$ = SURFACE AREA	1.742 in^2	11.24 cm^2
PERMEABILITY		60
μ 125		2.08 x L @ μ 60
μ 160		2.67 x L @ μ 60
μ 200		3.33 x L @ μ 60
μ 550		9.17 x L @ μ 60

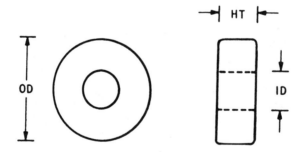

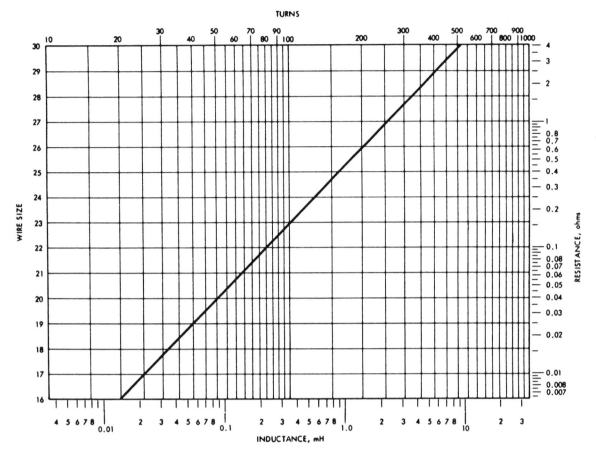

Figure 5.7. Wire and inductance graph for core 55121-A2.

Table 5.6. Dimensional specifications for Magnetics Inc. 55848-A2 and Arnold Engineering A-848032-2

	ENGLISH		METRIC	
Wa/Ac			4.91	
Wa x Ac	0.00636	in^4	0.264	cm^4
OD	0.830	in	2.11	cm
ID	0.475	in	1.21	cm
HT	0.280	in	0.711	cm
Wa = WINDOW AREA	0.23×10^6	CIR-MIL	1.14	cm^2
Wa = EFFECTIVE	0.13290	in^2	0.858	cm^2
Ac = CROSS SECTION	0.036	in^2	0.232	cm^2
Im = PATH LENGTH	2.01	in	5.09	cm
CORE WEIGHT	0.021	lb	9.6	grams
TOTAL WEIGHT	0.041	lb	18.6	grams
WOUND OD MIN	0.926	in	2.35	cm
MLT	1.166	in	2.97	cm
At = SURFACE AREA	2.431	in^2	15.69	cm^2
PERMEABILITY			60	
μ 125			2.08 x L @ μ 60	
μ 160			2.67 x L @ μ 60	
μ 200			3.33 x L @ μ 60	
μ 550			9.17 x L @ μ 60	

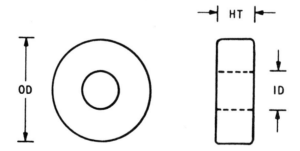

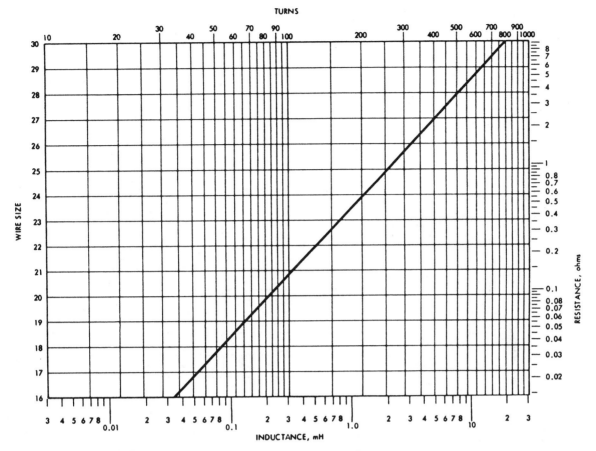

Figure 5.8. Wire and inductance graph for core 55848-A2.

Table 5.7. Dimensional specifications for Magnetics Inc. 55059-A2 and Arnold Engineering A-059043-2

	ENGLISH	METRIC
Wa/Ac		4.30
Wa x Ac	0.0713 in^4	0.460 cm^4
OD	0.930 in	2.36 cm
ID	0.527 in	1.339 cm
HT	0.330 in	0.838 cm
Wa = WINDOW AREA	0.28 x 10^6 CIR-MIL	1.407 cm^2
Wa = EFFECTIVE	0.164 in^2	1.056 cm^2
Ac = CROSS SECTION	0.0507 in^2	0.327 cm^2
lm = PATH LENGTH	2.23 in	5.67 cm
CORE WEIGHT	0.033 lb	15.0 grams
TOTAL WEIGHT	0.0716 lb	32.5 grams
WOUND OD MIN	1.035 in	2.63 cm
MLT	1.356 in	3.45 cm
A$_t$ = SURFACE AREA	3.103 in^2	20.019 cm^2
PERMEABILITY		60
μ 125		2.08 x L @ μ 60
μ 160		2.67 x L @ μ 60
μ 200		3.33 x L @ μ 60
μ 550		9.17 x L @ μ 60

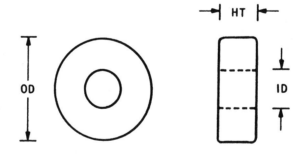

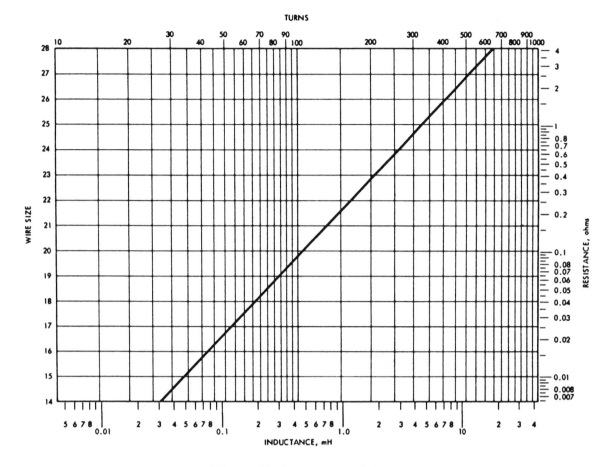

Figure 5.9. Wire and inductance graph for core 55059-A2.

Table 5.8. Dimensional specifications for Magnetics Inc. 55894-A2 and Arnold Engineering A-894075-2

	ENGLISH		METRIC	
Wa/Ac			2.44	
Wa x Ac	0.0239	in^4	0.997	cm^4
OD	1.090	in	2.77	cm
ID	0.555	in	1.41	cm
HT	0.472	in	1.20	cm
Wa = WINDOW AREA	0.31 x 10^6	CIR-MIL	1.561	cm^2
Wa = EFFECTIVE	0.1814	in^2	1.17	cm^2
Ac = CROSS SECTION	0.099	in^2	0.639	cm^2
lm = PATH LENGTH	2.50	in	6.35	cm
CORE WEIGHT	0.077	lb	35	grams
TOTAL WEIGHT	0.132	lb	59.7	grams
WOUND OD MIN	1.191	in	3.03	cm
MLT	1.81	in	4.61	cm
A$_t$ = SURFACE AREA	4.38	in^2	28.32	cm^2
PERMEABILITY			60	
μ 125			2.08 x L @ μ 60	
μ 160			2.67 x L @ μ 60	
μ 200			3.33 x L @ μ 60	
μ 550			9.17 x L @ μ 60	

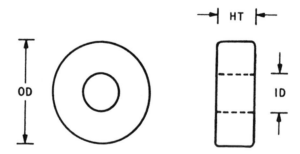

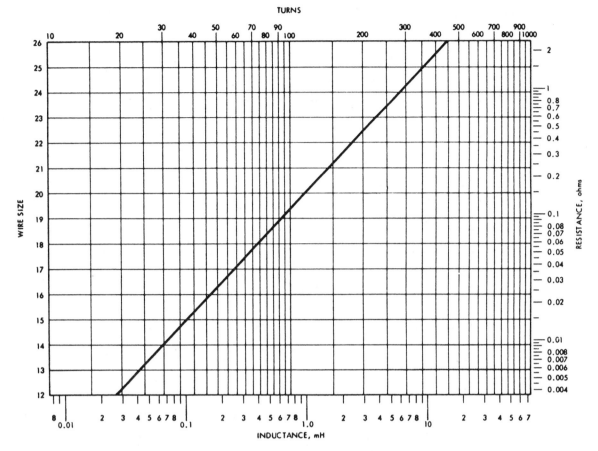

Figure 5.10. Wire and inductance graph for core 55894-A2.

Table 5.9. Dimensional specifications for Magnetics Inc. 55586-A2 and Arnold Engineering A3450387-2

	ENGLISH		METRIC	
W_a/A_c			8.73	
$W_a \times A_c$	0.044	in^4	1.832	cm^4
OD	1.382	in	3.51	cm
ID	0.888	in	2.26	cm
HT	0.387	in	0.983	cm
W_a = WINDOW AREA	0.79×10^6	CIR-MIL	4.00	cm^2
W_a = EFFECTIVE	0.4644	in^2	3.009	cm^2
A_c = CROSS SECTION	0.0710	in^2	0.458	cm^2
I_m = PATH LENGTH	3.53	in	8.95	cm
CORE WEIGHT	0.075	lb	34	grams
TOTAL WEIGHT	0.193	lb	87.4	grams
WOUND OD MIN	1.58	in	4.02	cm
MLT	1.70	in	4.32	cm
A_t = SURFACE AREA	6.85	in^2	44.24	cm^2
PERMEABILITY			60	
μ 125			$2.08 \times L @ \mu 60$	
μ 160			$2.67 \times L @ \mu 60$	
μ 200			$3.33 \times L @ \mu 60$	
μ 550			$9.17 \times L @ \mu 60$	

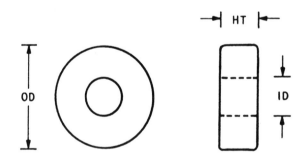

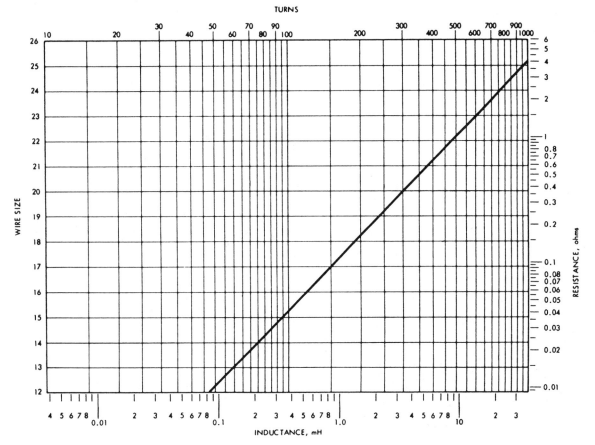

Figure 5.11. Wire and inductance graph for core 55586-A2.

Table 5.10. Dimensional specifications for Magnetics Inc. 55071-A2 and Arnold Engineering A-291061-2

	ENGLISH		METRIC	
Wa/Ac			4.39	
Wa x Ac	0.0468	in^4	1.95	cm^4
OD	1.332	in	3.38	cm
ID	0.760	in	1.93	cm
HT	0.457	in	1.16	cm
Wa = WINDOW AREA	0.58 x 10^6	CIR-MIL	2.93	cm^2
Wa = EFFECTIVE	0.340	in^2	2.1941	cm^2
Ac = CROSS SECTION	0.1032	in^2	0.666	cm^2
Im = PATH LENGTH	3.21	in	8.15	cm
CORE WEIGHT	0.101	lb	46	grams
TOTAL WEIGHT	0.198	lb	90	grams
WOUND OD MIN	1.486	in	3.77	cm
MLT	1.89	in	4.80	cm
A$_t$ = SURFACE AREA	4.389	in^2	40.68	cm^2
PERMEABILITY			60	
μ 125			2.08 x L @ μ 60	
μ 160			2.67 x L @ μ 60	
μ 200			3.33 x L @ μ 60	
μ 550			9.17 x L @ μ 60	

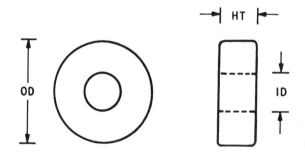

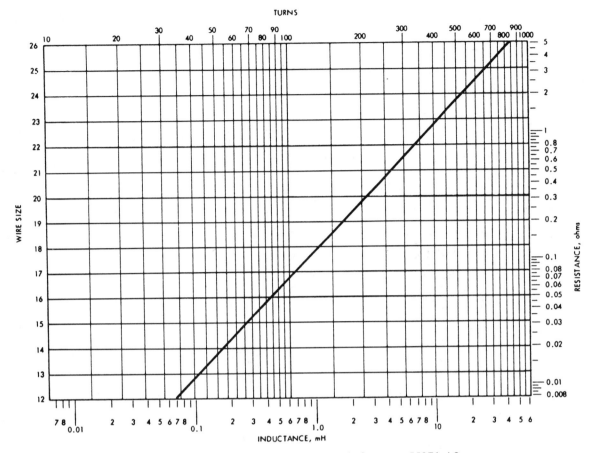

Figure 5.12. Wire and inductance graph for core 55071-A2.

Table 5.11. Dimensional specifications for Magnetics Inc. 55076-A2 and Arnold Engineering A-076056-2

	ENGLISH		METRIC	
Wa/Ac			5.43	
Wa x Ac	0.0586	in^4	2.44	cm^4
OD	1.44	in	3.66	cm
ID	0.848	in	2.15	cm
HT	0.444	in	1.128	cm
Wa = WINDOW AREA	0.72 x 10^6	CIR-MIL	3.64	cm^2
Wa = EFFECTIVE	0.424	in^2	2.723	cm^2
Ac = CROSS SECTION	0.1039	in^2	0.670	cm^2
Im = PATH LENGTH	3.54	in	8.98	cm
CORE WEIGHT	0.112	lb	51	grams
TOTAL WEIGHT	0.239	lb	108.4	grams
WOUND OD MIN	1.62	in	4.11	cm
MLT	1.91	in	4.88	cm
A$_t$ = SURFACE AREA	7.271	in^2	46.91	cm^2
PERMEABILITY			60	
μ 125			2.08 x L @ μ 60	
μ 160			2.67 x L @ μ 60	
μ 200			3.33 x L @ μ 60	
μ 550			9.17 x L @ μ 60	

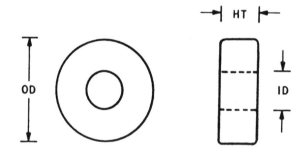

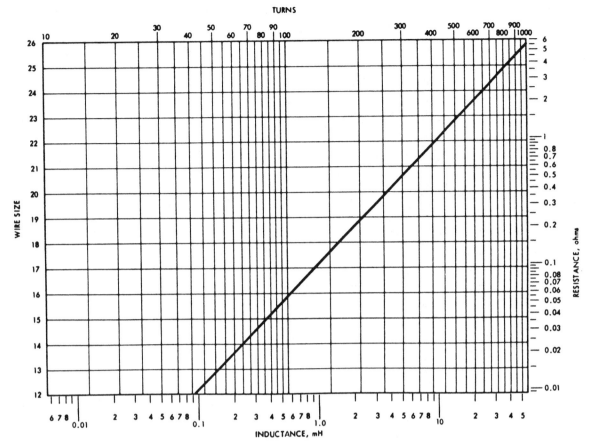

Figure 5.13. Wire and inductance graph for core 55076-A2.

Table 5.12. Dimensional specifications for Magnetics Inc. 55083-A2 and Arnold Engineering A-083081-2

	ENGLISH		METRIC	
W_a/A_c			4.02	
$W_a \times A_c$	0.108	in^4	4.53	cm^4
OD	1.602	in	4.07	cm
ID	0.918	in	2.33	cm
HT	0.605	in	1.54	cm
W_a = WINDOW AREA	0.84×10^6	CIR-MIL	4.27	cm^2
W_a = EFFECTIVE	0.496	in^2	3.198	cm^2
A_c = CROSS SECTION	0.164	in^2	1.06	cm^2
l_m = PATH LENGTH	3.88	in	9.84	cm
CORE WEIGHT	0.198	lb	90	grams
TOTAL WEIGHT	0.388	lb	176	grams
WOUND OD MIN	1.79	in	4.54	cm
MLT	2.36	in	6.07	cm
A_t = SURFACE AREA	9.46	in^2	61.05	cm^2
PERMEABILITY			60	
μ 125			$2.08 \times L$ @ μ 60	
μ 160			$2.67 \times L$ @ μ 60	
μ 200			$3.33 \times L$ @ μ 60	
μ 550			$9.17 \times L$ @ μ 60	

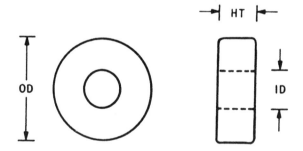

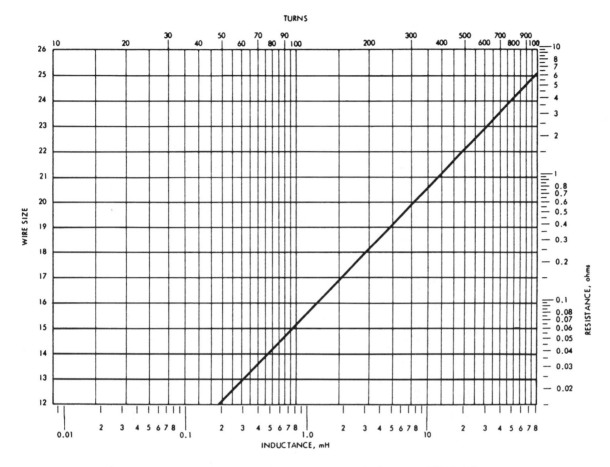

Figure 5.14. Wire and inductance graph for core 55083-A2.

Table 5.13. Dimensional specifications for Magnetics Inc. 55439-A2 and Arnold Engineering A-759135-2

	ENGLISH		METRIC	
Wa/Ac			2.19	
Wa x Ac	0.200	in^4	8.33	cm^4
OD	1.875	in	4.76	cm
ID	0.918	in	2.33	cm
HT	0.745	in	1.89	cm
Wa = WINDOW AREA	0.84×10^6	CIR-MIL	4.27	cm^2
Wa = EFFECTIVE	0.496	in^2	3.198	cm^2
Ac = CROSS SECTION	0.302	in^2	1.95	cm^2
lm = PATH LENGTH	4.23	in	10.74	cm
CORE WEIGHT	0.346	lb	180	grams
TOTAL WEIGHT	0.641	lb	291	grams
WOUND OD MIN	2.04	in	5.17	cm
MLT	3.00	in	7.62	cm
A_t = SURFACE AREA	12.30	in^2	79.37	cm^2
PERMEABILITY			60	
μ 125			2.08 x L @ μ 60	
μ 160			2.67 x L @ μ 60	
μ 200			3.33 x L @ μ 60	
μ 550			9.17 x L @ μ 60	

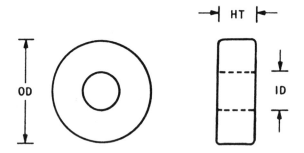

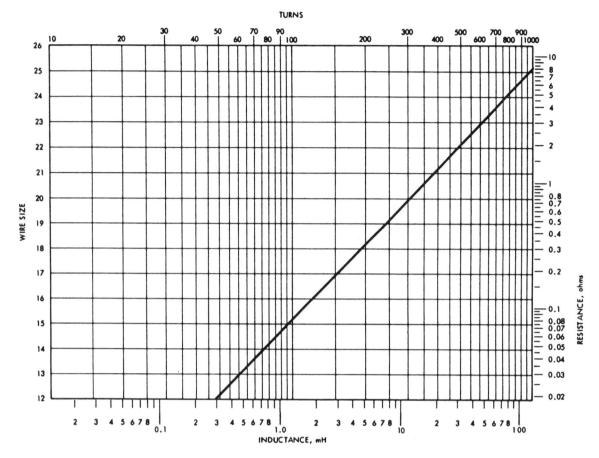

Figure 5.15. Wire and inductance graph for core 55439-A2.

Table 5.14. Dimensional specifications for Magnetics Inc. 55090-A2 and Arnold Engineering A-090086-2

	ENGLISH		METRIC	
W_a/A_c			4.63	
$W_a \times A_c$	0.194	in^4	8.06	cm^4
OD	1.875	in	4.76	cm
ID	1.098	in	2.79	cm
HT	0.635	in	1.61	cm
W_a = WINDOW AREA	1.21×10^6	CIR-MIL	6.11	cm^2
W_a = EFFECTIVE	0.710	in^2	4.58	cm^2
A_c = CROSS SECTION	0.205	in^2	1.32	cm^2
I_m = PATH LENGTH	4.58	in	11.62	cm
CORE WEIGHT	0.286	lb	130	grams
TOTAL WEIGHT	0.588	lb	267	grams
WOUND OD MIN	2.10	in	5.34	cm
MLT	2.62	in	6.66	cm
A_t = SURFACE AREA	12.64	in^2	81.58	cm^2
PERMEABILITY			60	
μ 125			2.08 x L @ μ 60	
μ 160			2.67 x L @ μ 60	
μ 200			3.33 x L @ μ 60	
μ 550			9.17 x L @ μ 60	

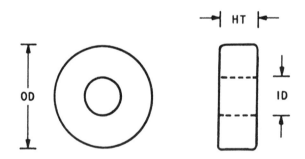

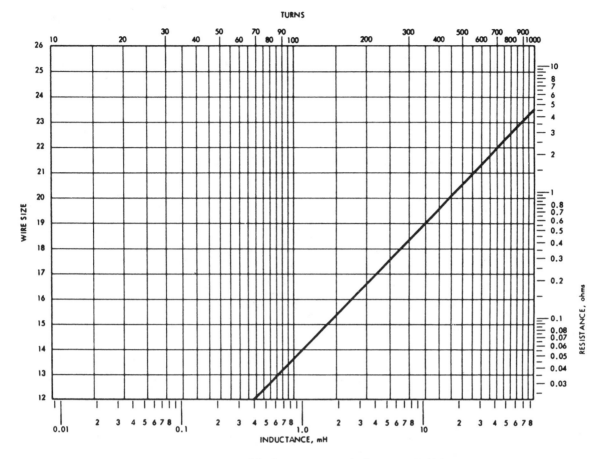

Figure 5.16. Wire and inductance graph for core 55090-A2.

Table 5.15. Dimensional specifications for Magnetics Inc. 55716-A2 and Arnold Engineering A-106073-2

	ENGLISH		METRIC	
Wa/Ac			6.06	
Wa x Ac	0.224	in^4	9.32	cm^4
OD	2.035	in	5.17	cm
ID	1.218	in	3.09	cm
HT	0.565	in	1.435	cm
Wa = WINDOW AREA	1.48 x 10^6	CIR-MIL	7.52	cm^2
Wa = EFFECTIVE	0.874	in^2	5.62	cm^2
Ac = CROSS SECTION	0.192	in^2	1.24	cm^2
Im = PATH LENGTH	5.02	in	12.73	cm
CORE WEIGHT	0.298	lb	135	grams
TOTAL WEIGHT	0.652	lb	296	grams
WOUND OD MIN	2.29	in	5.82	cm
MLT	2.55	in	6.50	cm
A$_t$ = SURFACE AREA	14.15	in^2	91.32	cm^2
PERMEABILITY			60	
μ 125			2.08 x L @ μ 60	
μ 160			2.67 x L @ μ 60	
μ 200			3.33 x L @ μ 60	
μ 550			9.17 x L @ μ 60	

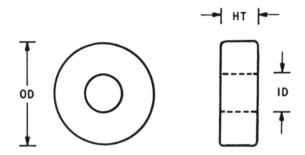

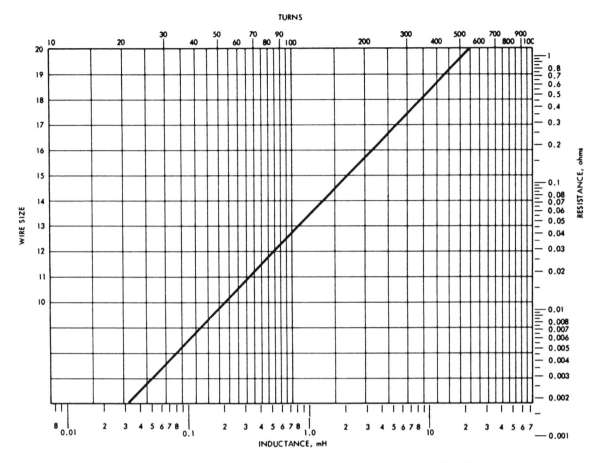

Figure 5.17. Wire and inductance graph for core 55716-A2.

Table 5.16. Dimensional specifications for Magnetics Inc. 55110-A2 and Arnold Engineering A-488075-2

	ENGLISH		METRIC	
Wa/Ac			6.58	
Wa x Ac	0.328	in^4	13.65	cm^4
OD	2.285	in	5.8	cm
ID	1.368	in	3.47	cm
HT	0.585	in	1.486	cm
Wa = WINDOW AREA	1.87 x 10^6	CIR-MIL	9.48	cm^2
Wa = EFFECTIVE	1.1023	in^2	7.093	cm^2
Ac = CROSS SECTION	0.223	in^2	1.44	cm^2
lm = PATH LENGTH	5.63	in	14.30	cm
CORE WEIGHT	0.385	lb	175	grams
TOTAL WEIGHT	0.864	lb	392	grams
WOUND OD MIN	2.57	in	6.53	cm
MLT	2.75	in	7.00	cm
A$_t$ = SURFACE AREA	17.42	in^2	112.4	cm^2
PERMEABILITY			60	
μ 125			2.08 x L @ μ 60	
μ 160			2.67 x L @ μ 60	
μ 200			3.33 x L @ μ 60	
μ 550			9.17 x L @ μ 60	

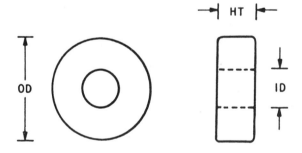

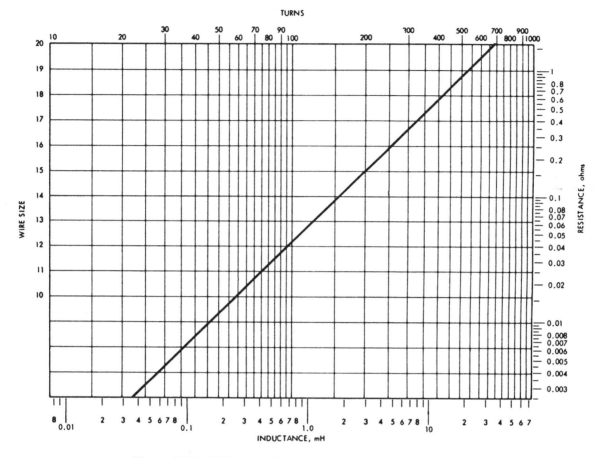

Figure 5.18. Wire and inductance graph for core 55110-A2.

Chapter 6

Window Utilization Factor K_u

6.1 INTRODUCTION

The window utilization factor K_u is a measure of the amount of copper that appears in the window area of the transformer or inductor. The window utilization factor is influenced by four factors:

1. Wire insulation
2. Wire lay (fill factor)
3. Bobbin area (or, when using a toroid, the clearance hole for passage of the shuttle)
4. Insulation required for multilayer windings or between windings

In the design of high-current or low-current transformers, the ratio of conductor area to total wire area can vary from 0.941 to 0.673, depending on the wire size. The wire lay or fill factor can vary from 0.7 to 0.55, depending on the winding technique. The amount and the type of insulation are dependent on the voltage.

6.2 THE WINDOW UTILIZATION FACTOR

The fraction K_u of the available core window space that will be occupied by the winding (copper) is calculated from areas S_1, S_2, S_3, and S_4:

$$K_u = S_1 \times S_2 \times S_3 \times S_4 \tag{6.1}$$

where

$$S_1 = \frac{\text{conductor area}}{\text{wire area}}$$

$$S_2 = \frac{\text{wound area}}{\text{usable window area}}$$

$$S_3 = \frac{\text{usable window area}}{\text{window area}}$$

$$S_4 = \frac{\text{usable window area}}{\text{usable window area} + \text{insulation area}}$$

in which

Conductor area = copper area

Wire area = copper area + insulation area

Wound area = number of turns × wire area of one turn

Usable window area = available window area − residual area that results from the particular winding technique used

Window area = available window area

Insulation area = area usable for winding insulation

S_1 is dependent upon wire size. Columns A and D of Table 6.1 may be used for calculating some typical values. For example,

$$S_1(\text{AWG 10}) = \frac{52.61 \text{ cm}^2}{55.90 \text{ cm}^2} = 0.941$$

$$S_1(\text{AWG 20}) = \frac{5.188 \text{ cm}^2}{6.065 \text{ cm}^2} = 0.855$$

$$S_1(\text{AWG } 30) = \frac{0.5067 \text{ cm}^2}{0.6785 \text{ cm}^2} = 0.747$$

$$S_1(\text{AWG } 40) = \frac{0.04869 \text{ cm}^2}{0.0723 \text{ cm}^2} = 0.673$$

When designing low-current transformers, it is advisable to reevaluate S_1 because of the increased amount of insulation.

S_2 is the fill factor for the usable window area. It can be shown that for wire of circular cross section wound on a flat form, the ratio of wire area to the area required for the turns can never be greater than 0.91. In practice, the actual maximum value depends upon the tightness of winding, variations in insulation thickness, and wire lay. Consequently, the fill factor is always less than the theoretical maximum.

As a typical working value for copper wire with a heavy synthetic film insulation, a ratio of 0.60 may be safely used for S_2.

The term S_3 defines how much of the available window space may actually be used for the winding. The winding area available to the designer depends on the bobbin configuration. A single-bobbin design offers an effective area W_a between 0.835 and 0.929, while a two-bobbin configuration offers an effective area W_a between 0.687 and 0.872. *A good value to use for S_3 both configurations is 0.75.* When designing with a pot core, S_3 has to be reduced, because the effective W_a varies between 0.55 and 0.71.

The term S_4 defines how much of the usable window space is actually being used for insulation. If the transformer has multiple secondaries with significant amounts of insulation, S_4 should be reduced by 10% for each additional secondary winding, partly because of the added space occupied by insulation and partly because of a poorer space factor.

A typical value for K_u, the copper fraction in the window area, is about 0.40. For example, for AWG 20 wire, $S_1 \times S_2 \times S_3 \times S_4 = 0.855 \times 0.60 \times 0.75 \times 1.0 = 0.385$, which is very close to 0.4. This may be stated somewhat differently as:

$$0.4 = \underbrace{\frac{A_w \text{ bare}}{A_w \text{ total}}}_{(S_1)} \times \underbrace{\text{fill factor}}_{(S_2)} \times \underbrace{\frac{W_{a(\text{eff})}}{W_a}}_{(S_3)} \times \underbrace{\text{insulation factor}}_{(S_4)}$$

6.3 CONVERSION DATA FOR WIRE SIZES No. 10 TO No. 44

Columns A and B in Table 6.1 give the bare area in the commonly used circular mils notation and in the metric equivalent for each wire size. Column C gives the equivalent resistance in micro-ohms per centimeter ($\mu\Omega$/cm, or 10^{-6} Ω/cm) in wire length for each wire size. Columns D–L relate to coated wires and show the effect of insulation on size, the number of turns, and total weight in grams per centimeter.

The total resistance for a given winding may be calculated by multiplying the MLT (mean length per turn) of the winding in centimeters by the micro-ohms per centimeter for the appropriate wire size (column C) and the total number of turns. Thus

Table 6.1. Wire table

AWG Wire Size	Bare Area		Resistance	Area		Diameter		Heavy Synthetics				Weight
	cm²10⁻³ (footnote b)	CIR-MIL[a]	10⁻⁶Ω cm at 20°C	cm²10⁻³	CIR-MIL[a]	cm	inch[a]	Turns per		Turns per		g/cm
								cm	inch[a]	cm²	inch²	
10	52.61	10384	32.70	55.9	11046	0.267	0.1051	3.87	9.5	10.73	69.20	0.468
11	41.68	8226	41.37	44.5	8798	0.238	0.0938	4.36	10.7	13.48	89.95	0.3750
12	33.08	6529	52.09	35.64	7022	0.213	0.0838	4.85	11.9	16.81	108.4	0.2977
13	26.26	5184	65.64	28.36	5610	0.190	0.0749	5.47	13.4	21.15	136.4	0.2367
14	20.82	4109	82.80	22.95	4556	0.171	0.0675	6.04	14.8	26.14	168.6	0.1879
15	16.51	3260	104.3	18.37	3624	0.153	0.0602	6.77	16.6	32.66	210.6	0.1492
16	13.07	2581	131.8	14.73	2905	0.137	0.0539	7.32	18.6	40.73	262.7	0.1184
17	10.39	2052	165.8	11.68	2323	0.122	0.0482	8.18	20.8	51.36	331.2	0.0943
18	8.228	1624	209.5	9.326	1857	0.109	0.0431	9.13	23.2	64.33	414.9	0.07472
19	6.531	1289	263.9	7.539	1490	0.0980	0.0386	10.19	25.9	79.85	515.0	0.05940
20	5.188	1024	332.3	6.065	1197	0.0879	0.0346	11.37	28.9	98.93	638.1	0.04726
21	4.116	812.3	418.9	4.837	954.8	0.0785	0.0309	12.75	32.4	124.0	799.8	0.03757
22	3.243	640.1	531.4	3.857	761.7	0.0701	0.0276	14.25	36.2	155.5	1003	0.02965
23	2.588	510.8	666.0	3.135	620.0	0.0632	0.0249	15.82	40.2	191.3	1234	0.02372
24	2.047	404.0	842.1	2.514	497.3	0.0566	0.0223	17.63	44.8	238.6	1539	0.01884
25	1.623	320.4	1062.0	2.002	396.0	0.0505	0.0199	19.80	50.3	299.7	1933	0.01498
26	1.280	252.8	1345.0	1.603	316.8	0.0452	0.0178	22.12	56.2	374.2	2414	0.01185
27	1.021	201.6	1687.6	1.313	259.2	0.0409	0.0161	24.44	62.1	456.9	2947	0.00945
28	0.8046	158.8	2142.7	1.0515	207.3	0.0366	0.0144	27.32	69.4	570.6	3680	0.00747
29	0.6470	127.7	2664.3	0.8548	169.0	0.0330	0.0130	30.27	76.9	701.9	4527	0.00602
30	0.5067	100.0	3402.2	0.6785	134.5	0.0294	0.0116	33.93	86.2	884.3	5703	0.00472
31	0.4013	79.21	4294.6	0.5596	110.2	0.0267	0.0105	37.48	95.2	1072	6914	0.00372
32	0.3242	64.00	5314.9	0.4559	90.25	0.0241	0.0095	41.45	105.3	1316	8488	0.00305
33	0.2554	50.41	6748.6	0.3662	72.25	0.0216	0.0085	46.33	117.7	1638	10565	0.00241
34	0.2011	39.69	8572.8	0.2863	56.25	0.0191	0.0075	52.48	133.3	2095	13512	0.00189
35	0.1589	31.36	10849	0.2268	44.89	0.0170	0.0067	58.77	149.3	2545	17060	0.00150
36	0.1266	25.00	13608	0.1813	36.00	0.0152	0.0060	65.62	166.7	3309	21343	0.00119
37	0.1026	20.25	16801	0.1538	30.25	0.0140	0.0055	71.57	181.8	3901	25161	0.000977
38	0.08107	16.00	21266	0.1207	24.01	0.0124	0.0049	80.35	204.1	4971	32062	0.000773
39	0.06207	12.25	27775	0.0932	18.49	0.0109	0.0043	91.57	232.6	6437	41518	0.000593
40	0.04869	9.61	35400	0.0723	14.44	0.0096	0.0038	103.6	263.2	8298	53522	0.000464
41	0.03972	7.84	43405	0.0584	11.56	0.00863	0.0034	115.7	294.1	10273	66260	0.000379
42	0.03166	6.25	55428	0.04558	9.00	0.00762	0.0030	131.2	333.3	13163	84901	0.000299
43	0.02452	4.84	70308	0.03683	7.29	0.00685	0.0027	145.8	370.4	16291	105076	0.000233
44	0.0202	4.00	85072	0.03165	6.25	0.00635	0.0025	157.4	400.0	18957	122272	0.000195
	A	B	C	D	E	F	G	H	I	J	K	L

[a]These data from REA Magnetic Wire Datalator (Ref. 1).
[b]This notation means that the entry in the column must be multiplied by 10⁻³

$$R = MLT \times N \times (\mu\Omega/cm) \times \zeta \times 10^{-6} \quad [\Omega] \tag{6.2}$$

Column C values are for 20°C. For a higher and lower temperature, use the resistance correction factor ζ (zeta) obtained from Figure 6.1. (Note that for a temperature of 20°C, $\zeta = 1.0$.)

The weight of the copper in a given winding may be calculated by multiplying the MLT by the g/cm (column L) and the total number of turns. Thus

$$W_t = MLT \times N \times (\text{column L}) \quad [g] \tag{6.3}$$

Turns per square inch and turns per square centimeter are based on a 60% wire fill factor. Mean length per turn for a given winding may be calculated with the aid of Figure 6.2. Figure 6.3 shows a

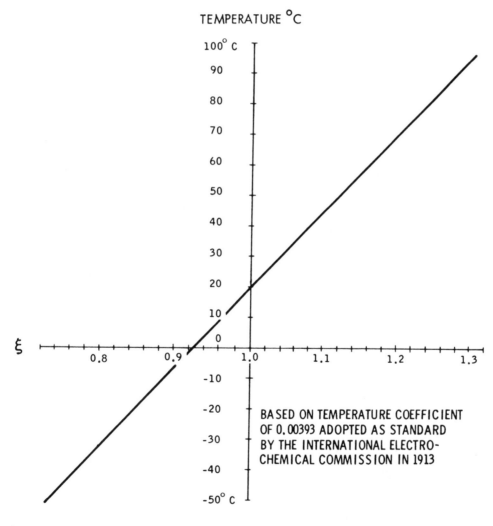

Figure 6.1. Resistance correction factor ζ (zeta) for wire resistance at temperatures between $-50°$ and 100°C.

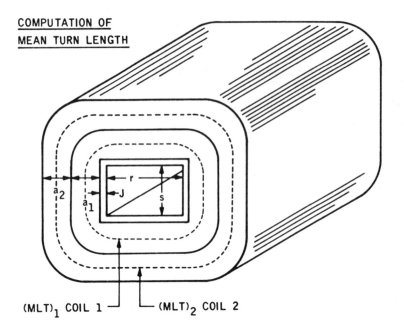

COMPUTATION OF
MEAN TURN LENGTH

(MLT)$_1$ COIL 1 —   — (MLT)$_2$ COIL 2

$$(MLT)_1 = 2(r+2J) + 2(s+2J) + \pi a_1$$
$$(MLT)_2 = 2(r+2J) + 2(s+2J) + \pi(2a_1+a_2)$$
OR
$$(MLT)_2 = (MLT)_1 + (a_1+a_2+2c)$$
OR
$$(MLT)_n = 2(r+2J) + 2(s+2J) + \pi\left[2(a_1+a_2+\ldots+a_{n-1}) + a_n\right]$$

WHERE:

a_1 = BUILD OF WINDING #1
a_2 = BUILD OF WINDING #2
a_n = BUILD OF WINDING #n
c = THICKNESS OF INSULATION BETWEEN a_1 & a_2

Figure 6.2. Computation of mean turn length (MLT).

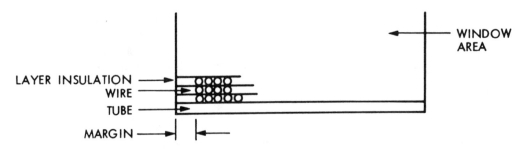

LAYER INSULATION →
WIRE →
TUBE →

MARGIN →

WINDOW
AREA

Figure 6.3. Layer-insulated coil.

transformer being constructed using layer insulation. When a transformer is being built in this way, Tables 6.2 and 6.3 will help the designer find the correct insulation thickness and margin for the appropriate wire size.

Table 6.2. Layer insulation vs. AWG

AWG	Insulation thickness	
	cm	inch
10-16	0.0254	0.01
17-19	0.0178	0.007
20-21	0.0127	0.005
22-23	0.0076	0.003
24-27	0.0051	0.002
28-33	0.00381	0.0015
34-41	0.00254	0.001
42-46	0.00127	0.0005

Table 6.3. Margin vs. AWG

AWG	Margin	
	cm	inch
10-15	0.635	0.250
16-18	0.475	0.187
19-21	0.396	0.156
22-31	0.318	0.125
32-37	0.236	0.093
38 →	0.157	0.062

6.4 TEMPERATURE CORRECTION FACTORS

The resistance values given in Table 6.1 are based on a temperature of 20°C. For other temperatures the effect upon wire resistance can be calculated by multiplying the resistance value for the wire size shown in column C of Table 6.1 by the appropriate correction factor shown on the graph of Figure 6.1. Thus,

$$\text{Corrected resistance} = \mu\Omega/\text{cm (at 20°C)} \times \zeta$$

6.5 WINDOW UTILIZATION FACTOR FOR A TOROID

The toroidal magnetic component has found wide use in industry and aerospace because of its high-frequency capability. The high-frequency capability of the toroid is due to its high ratio of window area to core cross section and its ability to accommodate different strip thicknesses in its boxed configuration. Tape strip thickness is an important consideration in selecting cores. Eddy current losses in the core can be reduced at higher frequencies by use of thinner strip stock. The high ratio of window area to core cross section ensures the minimum of iron and a large winding area to minimize the flux density and core loss.

The magnetic flux in the tape-wound toroid can be contained inside the core more readily than in lamination or C cores, as the winding covers the core along the entire magnetic path length, which gives lower electromagnetic interference.

The toroid does not give as smooth an A_p relationship as laminations, C cores, powder cores, and pot cores with respect to volume, weight, surface area, or current density, as can be seen in Chapter 2. This is because the actual core is always embedded in a case whose wall thickness has no fixed relation to the actual core and becomes relatively larger as the actual core cross section increases. The available window area inside the case, therefore, is not a fixed percentage of the window area of the uncased core.

Design Manual TWC-300 of Magnetics Inc. indicates that random-wound cores can be produced with fill factors as high as 0.7 but that progressive sector-wound cores can be produced with fill factors of only up to 0.55. As a typical working value for copper wire with a heavy synthetic film insulation, a ratio of 0.60 may be used safely. Figure 6.4 is based on a fill factor ratio of 0.60 for wire sizes 14–42 with 0.5 I.D. remaining.

Term S_3, the ratio of usable window area (cm^2) to window area (cm^2), defines how much of the available window space can actually be used for the winding. Figure 6.5 is based on the assumption that the inside diameter (I.D.) of the wound core is one-half that of the bare core, i.e., $S_3 = 0.75$ (to allow free passage of the shuttle).

Insulation factor S_4 in Figure 6.4 is 1.0; this does not take into account any insulation. The window utilization factor K_u is highly influenced by insulation factor S_4 because of the rapid buildup of insulation in a toroid as shown in Figure 6.6.

It can be seen in Figure 6.6 that the insulation buildup is greater on the inside than on the outside. For an example in Figure 6.6, if 1.27-cm wide tape were used with an overlap of 0.32 cm on the outside diameter, the overlap thickness would be four times the thickness of the tape. It will be noted that the amount of overlap will depend greatly on the size of the toroid. As the toroid window

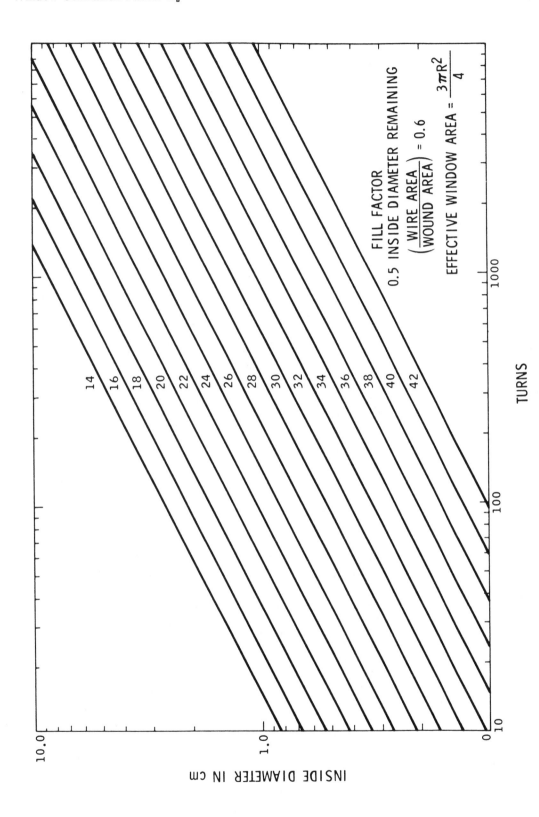

Figure 6.4. Toroid inside diameter vs. turns.

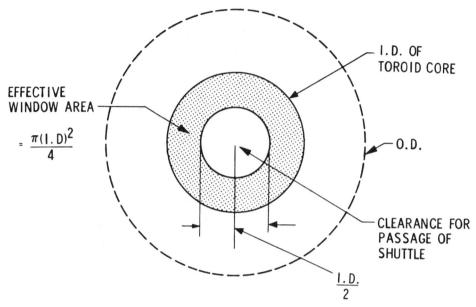

EFFECTIVE
WINDOW AREA

$= \dfrac{\pi (I.D)^2}{4}$

I.D. OF
TOROID CORE

O.D.

CLEARANCE FOR
PASSAGE OF
SHUTTLE

$\dfrac{I.D.}{2}$

Figure 6.5.   Effective winding area of a toroid.

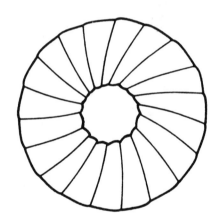

Figure 6.6.   Wrapped toroid.

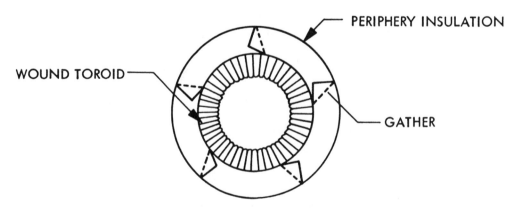

PERIPHERY INSULATION

WOUND TOROID

GATHER

Figure 6.7.   Periphery insulation.

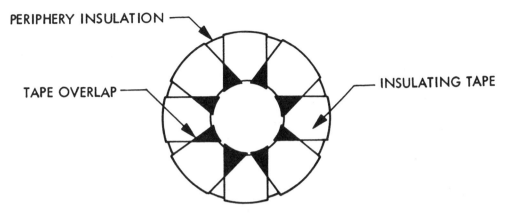

Figure 6.8. Minimizing toroidal inside build.

gets smaller, the overlap increases. There is a way to minimize the build on a wrapped toroid, and that is to use periphery insulation as shown in Figure 6.7. The use of periphery insulation minimizes the inside diameter overlay as shown in Figure 6.8.

When a design requires a multitude of windings, all of which have to be insulated, then the value of the insulation factor S_4 becomes very important in evaluating the window utilization factor K_u. For example, a low-current toroidal transformer with insulation has a significant influence on the window utilization factor as shown below for No. 40 AWG:

$$K_u = S_1 \times S_2 \times S_3 \times S_4$$

$$K_u = 0.673 \times 0.60 \times 0.75 \times 0.80$$

$$K_u = 0.242$$

Table 6.4 was generated as an aid for the engineer. It lists the metric dimensions weight, and area product A_p for 29 A.I.E.E. preferred tape-wound toroidal cores.

6.6 LEAKAGE INDUCTANCE

Operation of transformers at high frequency presents unique design problems, owing to the increased importance of core loss and leakage inductance. High-frequency designs therefore require considerably more care in specifying the type of core and windings, because the physical orientation and spacing of the winding determine leakage inductance. Leakage inductance is actually distributed throughout the windings in a transformer, because flux set up by the primary winding, which does not link the secondary, gives rise to leakage inductance in each winding without contributing to the mutual flux, as shown in Figure 6.9. However, for simplicity, it is shown as a lumped constant in Figure 6.10, where the leakage inductance is represented by L_p for the primary and L_s

Table 6.4. Dimensions, weight, and area product for A.I.E.E. preferred tape-wound toroidal cores

Mag Inc	Arnold	(1) A_c, cm^2	W_a, cm^2	(2) I.D., cm	O.D., cm	Ht, cm	I_m, cm	(3) Core wt, g	A_p, cm^4
52056	8T8043	0.043	0.915	1.079	1.778	0.559	4.49	1.67	0.0393
52000	8T5340	0.086	0.915	1.079	2.095	0.559	4.99	3.73	0.0787
52076	8T5958	0.193	1.478	1.372	2.756	0.711	6.48	10.9	0.285
52007	8T5651	0.257	1.478	1.372	2.756	0.876	6.19	14.5	0.380
52002	8T5515	0.086	1.674	1.460	2.476	0.559	6.98	4.62	0.144
52061	8T5502	0.171	2.274	1.702	2.743	0.876	7.48	10.4	0.389
52106	8T5504	0.193	2.274	1.702	3.061	0.711	8.98	12.6	0.439
52011	8T4168	0.086	4.242	2.324	3.391	0.559	8.98	6.71	0.365
52004	8T7699	0.171	4.242	2.324	3.391	0.876	9.43	13.4	0.725
52029	8T4635	0.257	4.242	2.324	3.701	0.876	9.97	21.2	1.090
52032	8T5800	0.343	4.242	2.324	4.026	0.876	9.97	29.8	1.455
52026	8T5233	0.514	4.242	2.324	4.026	1.194	9.97	44.7	2.180
52038	8T6847	0.686	4.242	2.324	4.026	1.537	11.96	59.6	2.910
52030	8T5387	0.343	6.816	2.946	4.674	0.889	11.96	35.8	2.379
52035	8T7441	0.686	6.816	2.946	4.674	1.549	11.96	65.6	4.676
52425	8T5772	0.771	6.816	2.946	5.308	1.219	12.96	87.2	5.255
52001	8T5320	1.371	9.648	3.505	6.629	1.575	15.95	191	13.23
52018	8T4179	0.257	11.55	3.835	5.372	0.876	14.46	32.4	2.968
52017	8T4178	0.686	18.19	4.813	6.617	1.575	17.95	107	12.48
52103	8T6110	1.371	17.91	4.775	7.925	1.587	19.94	238	24.55
52022	8T8027	2.742	17.91	4.775	7.925	2.845	19.94	477	49.11
52031	8T4180	0.686	28.22	5.994	7.899	1.575	21.93	131	19.36
52128	8T6100	1.371	28.22	5.994	9.195	1.613	23.93	286	38.68
52042	8T5468	2.742	28.22	5.994	9.195	2.883	23.93	572	77.38
52100	8T5690	5.142	28.22	5.994	9.881	4.216	24.93	1117	145.0
52081	8T5737	5.142	48.69	7.874	11.811	4.242	30.91	1386	250.3
52427	8T9259	7.198	48.37	7.848	13.106	4.305	32.90	2065	348.1
52112	8T5611	6.855	75.52	9.741	13.754	5.601	36.89	2205	517.7
52426	8T9260	10.968	74.14	9.716	15.680	5.601	39.88	3814	813.2

(1) Cross-sectional area calculated for 2 mil (0.002 in.) material
(2) Dimensions listed are sizes of aluminum boxed cores (not coated)
(3) 0.002 mil thickness and high nickel material

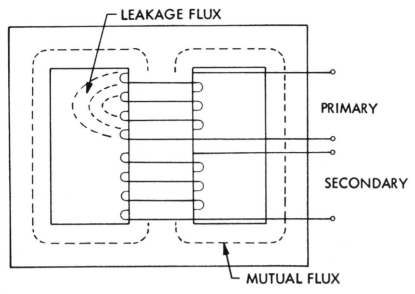

Figure 6.9. Leakage flux.

for the secondary. R_p and R_s are the equivalent dc resistance of the primary and secondary windings, respectively, R_e is the equivalent core-loss shunt resistance, and R_o is the load resistance.

Energy that is stored in the leakage flux can be detrimental in transistor switching circuits because when it is released it causes ringing and transistor failure. The value of the leakage inductance is almost independent of core material. Leakage inductance can be minimized by employing wide flat windings with a minimum of insulation; toroids provide a very low leakage inductance when the winding covers the entire magnetic path.

6.7 MEANS OF REDUCING LEAKAGE INDUCTANCE

There are a number of ways to reduce leakage inductance:

1. Minimize turns.
2. Reduce build of coil.

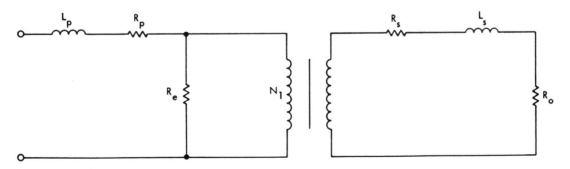

Figure 6.10. Lumped-constant equivalent circuit.

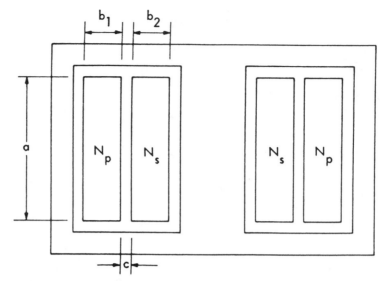

Figure 6.11. Conventional transformer configuration. (From R. Landee, *Electronic Designers Handbook.* Copyright 1975. Used with permission of McGraw-Hill Book Company.)

3. Increase winding width.
4. Minimize insulation between windings.
5. Use bifilar windings.

In the layer-wound coil, a substantial reduction in L_p and L_s is obtained by interweaving the primary and secondary windings. Figures 6.11–6.15 [see Equations (6.4)–(6.8), respectively] show winding configurations with the applicable equations for determining the total transformer leakage inductance as seen at the terminals of the primary winding. In Equations (6.4)–(6.8),

L_p is the primary leakage inductance, henrys
MLT is the mean length turn for whole coil, cm
N_p is the number of primary turns
a is the winding width, cm
b is the winding depth, cm
c is the dielectric thickness between windings, cm

$$L_p = \frac{1.2\,\text{MLT}\,N_p^2}{a}\left(c + \frac{b_1 + b_2}{3}\right) \times 10^{-8} \quad [\text{H}] \tag{6.4}$$

$$L_p = \frac{0.32\,\text{MLT}\,N_p^2}{a}\left(\Sigma c + \frac{\Sigma b}{3}\right) \times 10^{-8} \quad [\text{H}] \tag{6.5}$$

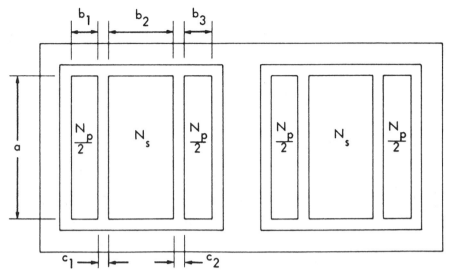

Figure 6.12. Sectionalizing the primary winding. (From R. Landee, *Electronic Designers Handbook*. Copyright 1975. Used with permission of McGraw-Hill Book Company.)

$$L_p = \frac{0.32\ \text{MLT}\ N_p^2}{a}\left(\Sigma c + \frac{\Sigma b}{3}\right) \times 10^{-8}\quad [\text{H}] \tag{6.6}$$

$$L_p = \frac{1.33\ \text{MLT}\ N_p^2}{b}\left(c + \frac{\Sigma a}{3}\right) \times 10^{-8}\quad [\text{H}] \tag{6.7}$$

$$L_p = \frac{0.37\ \text{MLT}\ N_p^2}{b}\left(\Sigma c + \frac{\Sigma a}{3}\right) \times 10^{-8}\quad [\text{H}] \tag{6.8}$$

The total leakage inductance referred to the secondary side is obtained by replacing N_p by N_s, where N_s is the number of secondary turns in any of Equations (6.4)–(6.8). It is quite difficult to reduce the leakage inductance that occurs in high-voltage transformers because of the large amount of insulation required.

6.8 MEASUREMENT OF LEAKAGE INDUCTANCE

Test sample will be the example in Chapter 11, "Three-Phase Transformer Design."

The secondary of the transformer is shorted, and then an ac signal is applied to the primary. Record the input current and voltage, then vice versa, as shown in Chapter 12, Figure 12.7. Use Equations (12.6) and (12.7) to solve for the inductance with the following values:

$$\text{MLT} = 13.4\quad [\text{cm}]$$

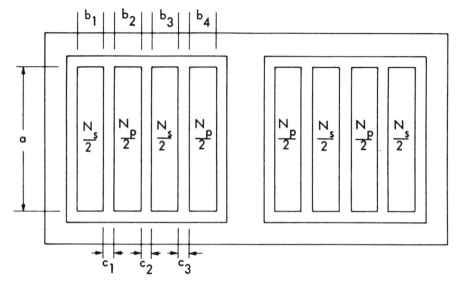

Figure 6.13. Sectionalizing the primary and secondary winding.

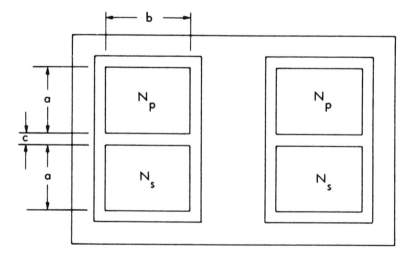

Figure 6.14. Sectionalizing with a two-section bobbin.

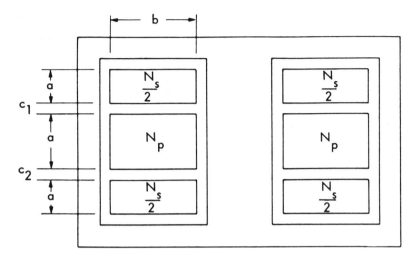

Figure 6.15. Sectionalizing with a three-section bobbin.

$$N_p = 1600$$

$$N_s = 100$$

$$a = 5.8 \quad [cm]$$

$$b_1 + b_2 = 1.15 \quad [cm]$$

$$c = 0.012 \quad [cm]$$

The test results, using Equation (6.4) for leakage inductance, are given in Table 6.5.

Table 6.5. Transformer leakage inductance

Item	Calculated Inductance henry	Measured Inductance henry
L_p	0.0295	0.032
L_s	0.000109	0.00012

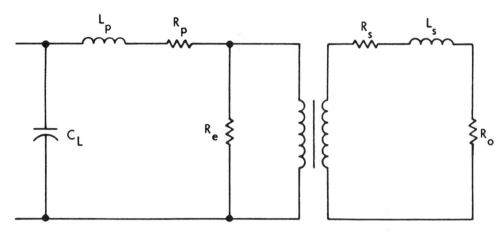

Figure 6.16. Transformer lumped capacitance.

6.9 WINDING CAPACITANCE

Operating at high frequency presents unique problems in the design of transformers to minimize the effects of winding capacitance. The capacitance C_L shown in Figure 6.16 represents the combined lumped capacitance of the primary and the secondary windings.

6.10 DISTRIBUTED CAPACITANCE

There are many capacitances that exist in a typical transformer, and they cannot be readily expressed in a single equation that would be useful for design purposes. However, the most important are these four:

1. Winding to core
2. Winding to winding
3. Layer to layer
4. Turn to turn

Their effect on transformer design varies with the application. The equation for calculating all these values will be given in the following test.

The real winding capacitance of a transformer as seen at the primary leads is almost impossible to measure accurately on a standard bridge, because the capacitance depends upon the actual voltage at the various points of the winding. The capacitance C_c between the first winding layer and the core is shown in Figure 6.17 along with the applicable equation. Equation (6.19) is also applicable to the capacitance between winding and winding.

In the multilayer winding shown in Figure 6.18, the effective layer-to-layer primary capacitance of the whole winding is expressed in Equation (6.10). The effective layer-to-layer secondary capacitance of the whole winding is done in the same manner with appropriate turns and insulation. The secondary capacitance reflected to the primary is shown in Figure 6.19.

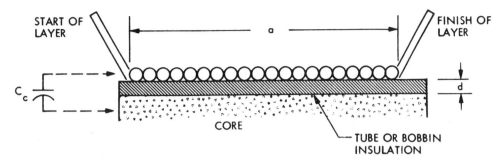

Figure 6.17. Winding to core capacitance.

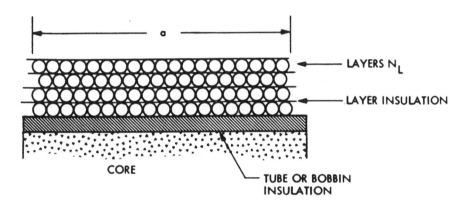

Figure 6.18. Multilayer winding capacitance.

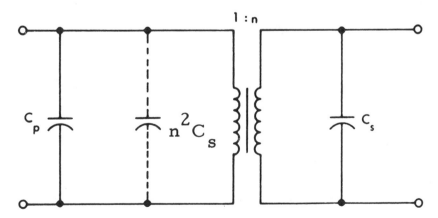

Figure 6.19. Secondary capacitance reflected to the primary.

$$C_p = \frac{4C_c}{3N_L}\left(1 - \frac{1}{N_L}\right) \quad (\text{pF}) \tag{6.10}$$

$$N_L = \text{number of layers}$$

For the most part, the turn-to-turn capacitance is small, because these capacitances are in series when referred to the whole winding. This capacitance is usually important only at 5 mc and above.

The lumped capacitance can then be expressed with the equation

$$C_L = C_p + n^2 C_s \quad [\text{pF}] \tag{6.11}$$

6.11 MEANS OF MINIMIZING CAPACITANCE

There are six major ways to minimize capacitance:

1. Increase the dielectric thickness d.
2. Reduce the winding width a.
3. Increase the number of layers.
4. Provide a large potential difference between winding.
5. Do not bifilar wind.
6. Use a Faraday (electrostatic) shield.

Tradeoffs have to be made in minimizing transformer capacitance and leakage inductance because corrective measures work in opposite directions.

In transformers designed to operate with a square wave such as a dc-to-dc converter, leakage inductance L_p and lumped capacitance C_L should be kept to a minimum, because they cause overshoot and oscillation or ringing, as shown in Figure 6.20.

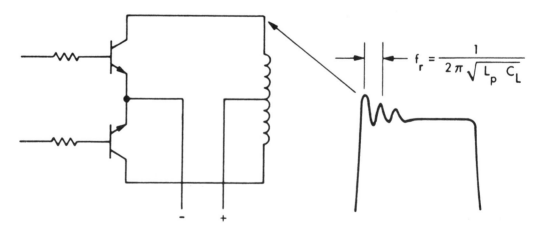

$$f_r = \frac{1}{2\pi\sqrt{L_p C_L}}$$

Figure 6.20. Transformer ringing.

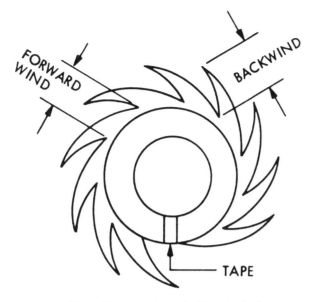

Figure 6.21. Progressive winding technique.

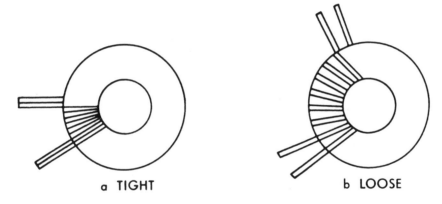

Figure 6.22. Minimizing capacitance effect on bifilar windings.

One way to reduce the capacitance effect in the transformer is to wind progressively and insulate the ends with tape as shown in Figure 6.21.

Some transformers require bifilar windings because of exact turns or dc resistance. This type of winding increases the capacitance quite rapidly. To minimize this effect, spread the wires as shown in Figure 6.22.

BIBLIOGRAPHY

Geossner, N., *Transformer for Electronic Circuits.* McGraw-Hill, New York, 1967.

Landee, R., Davis, D., and Albecht, A., *Electronic Designer's Handbook,* McGraw-Hill, New York, 1957, p. 14–12.

Lee, R., *Electronic Transformer and Circuits,* 2nd ed., John Wiley & Sons, New York, 1958.

Reference Data for Radio Engineers, 4th ed., International Telephone and Telegraph Co., New York.

Richardson, I., *The Technique of Transformer Design,* Electro-Technology, January 1961, pp. 58–67.

Chapter 7

Transformer-Inductor Efficiency, Regulation, and Temperature Rise

7.1 INTRODUCTION

Transformer efficiency, regulation, and temperature rise are all interrelated. Not all of the input power to the transformer is delivered to the load. The difference between the input power and output power is converted into heat. This power loss can be broken down into two compartments: core loss and copper loss. The core loss is a fixed loss, and the copper loss is a variable loss that is related to the current demand of the load. Copper loss increases by the square of the current and is also termed *quadratic loss*. Maximum efficiency is achieved when the fixed loss is equal to the quadratic loss at the rated load. Transformer regulation is the copper loss P_{cu} divided by the output power P_o.

7.2 TRANSFORMER EFFICIENCY

The efficiency of a transformer is a good way to measure the effectiveness of the design. Efficiency is defined as the ratio of the output power P_o to the input power P_{in}. The difference between P_o and P_{in} is due to losses. The total power loss, P_Σ, in a transformer is determined by the fixed losses in the core and the quadratic losses in the windings or copper. Thus,

$$P_\Sigma = P_{fe} + P_{cu} \qquad (7.1)$$

where P_{fe} is the core loss and P_{cu} is the copper loss.

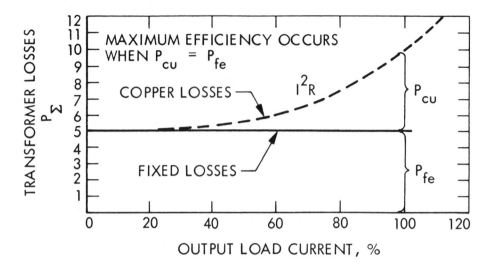

Figure 7.1. Transformer loss vs. output load current.

Maximum efficiency is achieved when the fixed loss is made equal to the quadratic loss as shown later by Equation (7.11). A graph of transformer loss versus output load current is shown in Figure 7.1.

The copper loss increases as the square of the output power multiplied by a constant K:

$$P_{cu} = KP_o^2 \tag{7.2}$$

which may be rewritten as

$$P_\Sigma = P_{fe} + KP_o^2 \tag{7.3}$$

Since

$$P_{in} = P_o + P_\Sigma \tag{7.4}$$

the efficiency can be expressed as

$$\eta = \frac{P_o}{P_o + P_\Sigma} \tag{7.5}$$

Then, substituting Equation (7.3) into (7.5) gives

$$\eta = \frac{P_o}{P_o + P_{\text{fe}} + KP_o^2} = \frac{P_o}{P_{\text{fe}} + P_o + KP_o^2} \tag{7.6}$$

and, differentiating with respect to P_o,

$$\frac{d\eta}{dP_o} = -P_o(P_{\text{fe}} + P_o + KP_o^2)^{-2}(1 + 2KP_o) \tag{7.7}$$

$$+ (P_{\text{fe}} + P_o + KP_o^2) = 0 \qquad \text{for max } \eta \tag{7.8}$$

$$-P_o(1 + 2KP_o) + (P_{\text{fe}} + P_o + KP_o^2) = 0 \tag{7.9}$$

$$-P_o - 2KP_o^2 + P_{\text{fe}} + P_o + KP_o^2 = 0 \tag{7.10}$$

Therefore,

$$P_{\text{fe}} = KP_o^2 = P_{\text{cu}} \tag{7.11}$$

7.3 RELATIONSHIP OF A_p TO CONTROL OF TEMPERATURE RISE

7.3.1 Temperature Rise

Not all of the input power P_{in} to the transformer is delivered to the load as P_o. Some of the input power is converted to heat by hysteresis and eddy currents induced in the core material and by the resistance of the windings. The first is a fixed loss arising from core excitation and is termed *core loss*. The second is a variable loss in the windings, which is related to the current demand of the load and thus varies as I^2R. This is termed the *quadratic* or *copper loss*.

The heat generated produces a temperature rise, which must be controlled to prevent damage to or failure of the windings by breakdown of the wire insulation at elevated temperatures. This heat is dissipated from the exposed surfaces of the transformer by a combination of radiation and convection. The dissipation is therefore dependent upon the total exposed surface area of the core and windings.

Ideally, maximum efficiency is achieved when the fixed and quadratic losses are equal. Thus,

$$P_\Sigma = P_{\text{fe}} + P_{\text{cu}} \tag{7.12}$$

and

$$P_{cu} = \frac{P_\Sigma}{2} \qquad (7.13)$$

When the copper loss in the primary winding is equal to the copper loss in the secondary, the current density in the primary is the same as the current density in the secondary:

$$\frac{P_p}{R_p} = \frac{P_s}{R_s} \qquad (7.14)$$

and

$$\frac{P_\Sigma}{R_t} = \frac{2P_p}{R_p/2} = \frac{4P_p}{R_p} = (2I_p)^2 \qquad (7.15)$$

Then

$$J_p = \frac{I_p}{W_a/2} = \frac{2I_p}{W_a} = J_s = J \qquad (7.16)$$

7.3.2. Calculation of Temperature Rise

Temperature rise in a transformer winding cannot be predicted with complete precision, despite the fact that many techniques are described in the literature for its calculation. One reasonably accurate method for open core and winding construction is based upon the assumption that core and winding losses may be lumped together as

$$P_\Sigma = P_{fe} + P_{cu} \qquad (7.17)$$

and the assumption is made that thermal energy is dissipated uniformly throughout the surface area of the core and winding assembly.

Transfer of heat by thermal *radiation* occurs when a body is raised to a temperature above its surroundings and emits radiant energy in the form of waves. In accordance with the Stefan-Boltzmann law,* this may be expressed as

$$W_r = K_r \varepsilon (T2^4 - T1^4) \qquad (7.18)$$

---

*See Reference 2, Chapter 3.

where W_r is watts per square centimeter of surface

    $K_r = 5.70 \times 10^{-12}$ W/(cm^2 = K^4)

     ε is the emissivity factor

    $T2$ is the hot body temperature, K

    $T1$ is the ambient or surrounding temperature, K

Transfer of heat by *convection* occurs when a body is hotter than the surrounding medium, which is usually air. The layer of air in contact with the hot body that is heated by conduction expands and rises, taking the absorbed heat with it. The next layer, being colder, replaces the risen layer and, in turn, on being heated also rises. This continues as long as the air or other medium surrounding the body is at a lower temperature. The transfer of heat by convection is stated mathematically as

$$W_c = K_c F \theta^{\eta} \sqrt{P} \tag{7.19}$$

where W_c is the watts loss per square centimeter

    $K_c = 2.17 \times 10^{-4}$

    F is the air friction factor (unity for a vertical surface)

    θ is the temperature rise, °C

    P is the relative barometric pressure (unity at sea level)

    η is the exponential value, which ranges from 1.0 to 1.25, depending on the shape and position of the surface being cooled

The total heat dissipated from a plane vertical surface is expressed by the sum of Equations (7.18) and (7.19):

$$W = 5.70 \times 10^{-12}\varepsilon(T2^4 - T1^4) + (1.4 \times 10^{-3})F\theta^{1.25}\sqrt{P} \tag{7.20}$$

7.3.3 Temperature Rise Versus Surface Area Dissipation

The temperature rise that can be expected for various levels of power loss is shown in the nomograph of Figure 7.2. It is based on Equation (7.20), relying on data obtained from Blyme (1938)* for heat transfer effected by a combination of 55% radiation and 45% convection, from surfaces having an emissivity of 0.95, in an ambient temperature of 25°C, at sea level. Power loss (heat dissipation) is expressed in watts per square centimeter of total surface area. Heat dissipation by convection from the upper side of a horizontal flat surface is on the order of 15–20% more than from vertical surfaces. Heat dissipation from the underside of a horizontal flat surface depends upon surface area and conductivity.

---

*See Reference 2, Chapter 3.

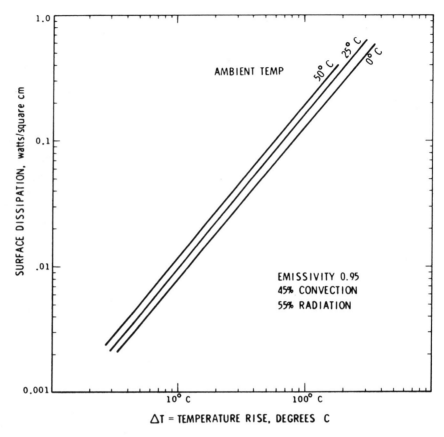

Figure 7.2. Temperature rise vs. surface dissipation. (Adapted from L. F. Blume, *Transformer Engineering,* Wiley, New York, 1938, Figure 7.)

7.3.4 Surface Area Required for Heat Dissipation

The effective surface area A_t required to dissipate heat (expressed as watts dissipated per unit area) is

$$A_t = \frac{P_\Sigma}{\psi} \tag{7.21}$$

in which ψ is the power density or the average power dissipated per unit area from the surface of the transformer and P_Σ is the total power lost or dissipated.

Surface area A_t of a transformer can be related to the area product A_p of a transformer. The straight-line logarithmic relationship shown in Figure 7.3 has been plotted from the data shown in Table 2.5, Chapter 2.

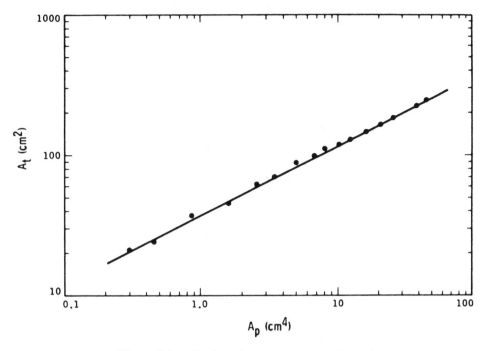

Figure 7.3. Surface area vs. area product A_p.

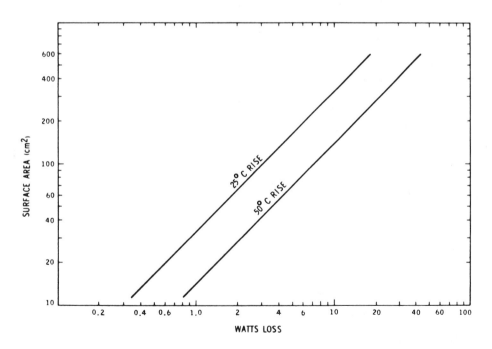

Figure 7.4. Surface area vs. total watt loss for temperature increases of 25°C and 50°C.

From this, the following relationship evolves:

$$A_t = K_s (A_p)^{0.5} = \frac{P_\Sigma}{\psi}$$ (7.22)

and (from Figure 7.2)

$$\psi = 0.03 \ W/cm^2 \text{ at } 25°C \text{ rise}$$ (7.23)

$$\psi = 0.07 \ W/cm^2 \text{ at } 50°C \text{ rise}$$ (7.24)

Figure 7.4 utilizes the efficiency rating in watts dissipated in terms of two different, but commonly allowable, temperature rises for the transformer over ambient temperature. The data presented are used as bases for determining the needed transformer surface area A_t (in cm^2).

7.4 REGULATION AS A FUNCTION OF EFFICIENCY

The minimum size of a transformer is usually determined either by a temperature rise limit or by allowable voltage regulation, assuming that size and weight are to be minimized.

Figure 7.5 shows the circuit diagram of a transformer with one secondary. The transformer window allocation is shown in Figure 7.6. This assumes that distributed capacitance in the secondary can be neglected because the frequency and secondary voltage are not excessively high. Also, the winding geometry is designed to limit the leakage inductance to a level low enough to be neglected under most operating conditions.

$$\frac{W_a}{2} = \text{Primary} = \text{Secondary}$$

Transformer voltage regulation can now be expressed as

$$\alpha = \frac{V_o(\text{NL}) - V_o(\text{FL})}{V_o(\text{FL})} \times 100$$ (7.25)

in which $V_o(\text{NL})$ is the no-load voltage and $V_o(\text{FL})$ is the full-load voltage.

Regulation for a two-winding isolation transformer can be expressed as

$$\alpha = \frac{\Delta V_p}{V_p}(100) + \frac{\Delta V_s}{V_s}(100)$$ (7.26)

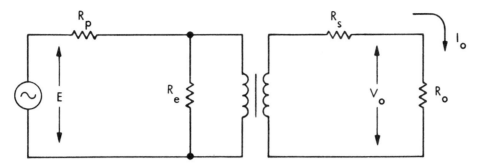

Figure 7.5.   Transformer circuit diagram.

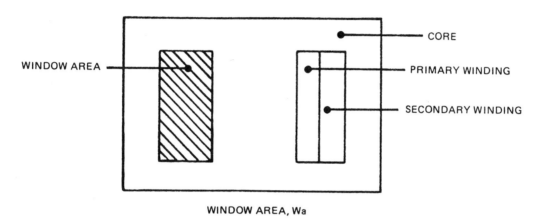

WINDOW AREA, Wa

Figure 7.6.   Transformer window allocation.

where

$$\Delta V_p = R_p I_p \tag{7.27}$$

$$\Delta V_s = R_s I_s \tag{7.28}$$

If the transformer had a 1:1 turns ratio, then

$$I_s \cong I_p = I_o$$

Multiplying the terms in Equation (7.25) by I_p/I_p and I_s/I_s, we get

$$\alpha = \frac{\Delta V_p I_p}{V_p I_p} \times 100 + \frac{\Delta V_s I_s}{V_s I_s} \times 100 \tag{7.29}$$

Then, since the primary copper loss is

$$P_p = \Delta V_p I_p \tag{7.30}$$

the secondary copper loss is

$$P_s = \Delta V_s I_s \tag{7.31}$$

and the total copper loss is

$$P_{cu} = P_p + P_s \tag{7.32}$$

Equation (7.29) can be expressed as

$$\alpha = \frac{P_{cu}}{P_o} \tag{7.33}$$

The input power is then:

$$P_{in} = P_{cu} + P_{fe} + P_o \tag{7.34}$$

For regulation as a function of efficiency,

$$\frac{P_o}{P_{in}} = \frac{P_o}{P_{cu} + P_{fe} + P_o} = \eta \tag{7.35}$$

By definition,

$$P_{cu} = P_{fe} \tag{7.36}$$

Solving for $P_{cu} + P_{fe}$:

$$\frac{P_o(1 - \eta)}{\eta} = P_o\left(\frac{1}{\eta} - 1\right) = P_{cu} + P_{fe} = 2P_{cu} \tag{7.37}$$

$$\frac{\alpha}{100} = \frac{1}{1 + P_o/P_{cu}} = \frac{1}{1 + 2/(1/\eta - 1)} = \frac{1 - \eta}{1 + \eta} \tag{7.38}$$

$$\alpha = \frac{1 - \eta}{1 + \eta} \times 100 \tag{7.39}$$

For efficiency as a function of regulation, multiply both sides of the equation by $(1 + \eta)$:

$$\alpha + \eta\alpha = 100 - \eta\, 100 \tag{7.40}$$

Solve for η:

$$\eta 100 + \eta\alpha = 100 - \alpha \tag{7.41}$$

$$\eta(100 + \alpha) = 100 - \alpha \tag{7.42}$$

$$\eta = \frac{100 - \alpha}{100 + \alpha} \tag{7.43}$$

7.5 MAGNETIC CORE MATERIAL TRADEOFF

The relationships between area product A_p and the core geometry coefficient K_g are associated only with such geometric properties as surface area and volume, weight, and the factors affecting temperature rise such as current density. A_p and K_g have no relevance to the magnetic core materials used. However, the designer often must make tradeoffs between such goals as efficiency and size, which are influenced by core material selection.

Usually in articles written about inverter and converter transformer design, recommendations with respect to choice of core material are a compromise of material characteristics such as those tabulated in Table 7.1 and graphically displayed in Figure 7.7. The characteristics shown here are those typical of commercially available core materials. As can be seen, the core material that provides the highest flux density is Supermendur. It also produces the smallest component size. If size is the most important consideration, this should determine the choice of materials. On the other hand, the type 78 Supermalloy material (see the 5/78 curve in Figure 7.7) has the lowest flux density. and this material would result in the largest size transformer. However, it also has the lowest coercive force and lowest core loss of any of the available materials. These factors might well be decisive in other applications. Choice of core material is thus based upon achieving the best characteristic for the most critical or important design parameter, with acceptable compromises on all other parameters.

Table 7.1. Magnetic core material characteristics

TRADE NAMES	COMPOSITION	* SATURATED FLUX DENSITY, tesla	DC COERCIVE FORCE, AMP–TURN / cm	SQUARENESS RATIO	** MATERIAL DENSITY, g / cm^3	CURIE TEMPERTURE, ^{0}C	WEIGHT FACTOR
Supermendur Permendur	49% Co 49% Fe 2% V	1.9–2.2	0.18–0.44	0.90–1.0	8.15	930	1.066
Magnesil Silectron Microsil Supersil	3% Si 97% Fe	1.5–1.8	0.5–0.75	0.85–0.75	7.63	750	1.00
Deltamax Orthonol 49 Sq Mu	50% Ni 50% Fe	1.4–1.6	0.125–0.25	0.94–1.0	8.24	500	1.079
Allegheny 4750 48 Alloy Carpenter 49	48% Ni 52% Fe	1.15–1.4	0.062–0.187	0.80–0.92	8.19	480	1.073
4-79 Permalloy Sp Permalloy 80 Sq Mu 79	79% Ni 17% Fe	0.66–0.82	0.025–0.82	0.80–1.0	8.73	460	1.144
Supermalloy	78% Ni 17% Fe 5% Mo	0.65–0.82	0.0037–0.01	0.40–0.70	8.76	400	1.148
Metglas: 2605SC	81% Fe 3.5% Si 13.5% B 2% C	1.5–1.6	.03–.08	.9–.98	7.32	370	.96
2714A	66% Co 4% Fe 15% Si 14% B 1% Ni	.5–.65	.008–.015	.9–.98	7.59	205	.995
Ferrites F N27 3C8	Mn Zn	0.45–0.50	0.25	0.30–0.5	4.8	250	0.629

*tesla = 10^4 Gauss
** g / cm^3 = 0.036 lb / in^3

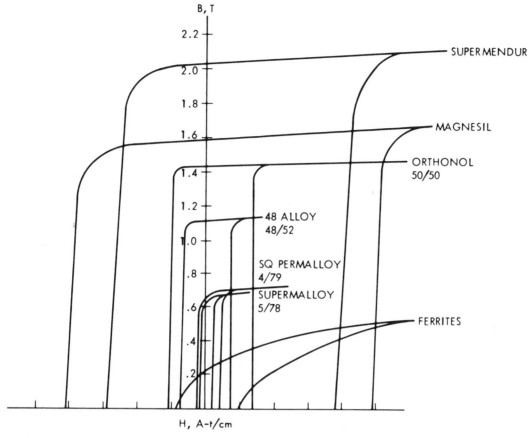

Figure 7.7. The typical dc *B-H* loops of magnetic material. (*Courtesy of Magnetics Inc.*)

7.6 SKIN EFFECT

It is now common practice to operate dc-to-dc converters at frequencies up to 300 kHz. At the higher frequencies, the predicted efficiency is altered, since the current carried by a conductor is distributed uniformly across the conductor cross section only with direct current and at low frequencies. There is a concentration of current near the wire surface at higher frequencies, which is termed the *skin effect*. This is the result of magnetic flux lines that circle only part of the conductor. Those portions of the cross section that are circled by the largest number of flux lines exhibit greater reactance.

Skin effect accounts for the fact that the ratio of effective alternating current resistance to direct current is greater than unity. The magnitude of this effect at high frequency on conductivity, magnetic permeability, and inductance is sufficient to require further evaluation of conductor size during design. The depth of the skin effect is expressed by

$$\text{Depth (cm)} = (6.61/f^{1/2})K \qquad (7.53)$$

in which K is the constant

$$K = \left[\left(\frac{1}{\mu_r}\right)\left(\frac{\rho}{\rho c}\right)\right]^{1/2} \qquad (7.54)$$

where μ_r is the relative permeability of conductor material ($\mu_r = 1$ for copper and other nonmagnetic materials)
ρ is the resistivity of conductor material at any temperature
c is the resistivity of copper; at 20°C, $c = 1.724$ μΩ-cm

For copper, $K = 1$.

Figure 7.8 is a graph of skin depth as a function of frequency according to Equation (7.53). The relationship of skin depth to AWG radius is illustrated in Figure 7.9, where $R_{ac}/R_{dc} = 1$ is plotted on a graph of AWG versus frequency.* Figure 7.10 shows how the rms values change with different waveshapes.

Figure 7.10 shows how the rms values change with different waveshapes.

---

*The data presented is for sine wave excitation. The author could not find any data for square wave excitation.

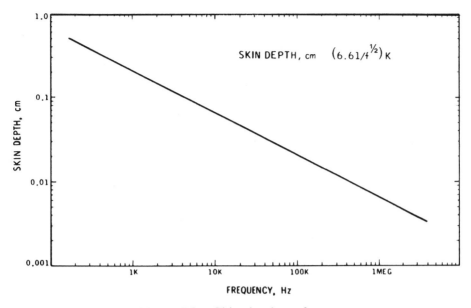

Figure 7.8. Skin depth vs. frequency.

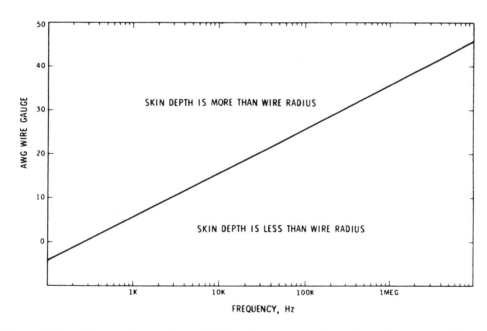

Figure 7.9. Skin depth equal to AWG radius versus plotted as a function of frequency.

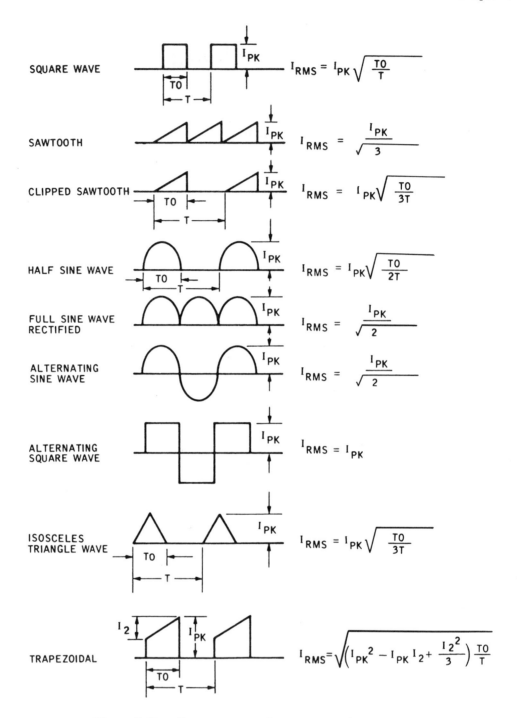

Figure 7.10. Common waveshapes, rms values.

Chapter 8

Inductor Design With No DC Flux

8.1 INTRODUCTION

The design of an ac inductor is quite similar to that of a transformer. If there is no dc flux in the core, the design calculations are straightforward.

The apparent power P_t of an inductor is the VA of the inductor; that is, the product of the excitation voltage and the current through the inductor:

$$P_t = VA \quad [\text{W}] \tag{8.1}$$

8.2 RELATIONSHIP OF A_p TO INDUCTOR VOLT-AMPERE CAPABILITY

The volt-ampere capability of a core is related to its area product A_p by an equation that may be stated as follows:

$$A_p = \left(\frac{VA \times 10^4}{4.44 B_m f K_u K_j} \right)^{1.14} \tag{8.2}$$

where K_j is the current density coefficient (see Chapter 2)
$\quad K_u$ is the window utilization factor (see Chapter 6)
$\quad f$ is the frequency, Hz
$\quad B_m$ is the flux density, T

From the above it can be seen that factors such as flux density, the window utilization factor K_u (which defines the maximum space that may be occupied by the copper in the window), and the constant K_j (which is related to temperature rise) all have an influence on the inductor area product. The constant K_j is a new parameter that gives the designer control of the copper loss. Derivation is set forth in detail in Chapter 2.

8.3 FUNDAMENTAL CONSIDERATIONS

The design of a linear inductor depends upon four related factors:

1. Desired inductance
2. Applied voltage
3. Frequency
4. Operating flux density

With these requirements established, the designer must determine the maximum values for B_{ac} that will not produce magnetic saturation, and make tradeoffs that will yield the highest inductance for a given volume. The core material selected determines the maximum flux density that can be tolerated for a given design. Magnetic saturation values for a number of core materials are given in Table 4.1.

The number of turns is calculated from the Faraday law, which states:

$$N = \frac{E \times 10^4}{4.44 B_m f A_c} \tag{8.3}$$

The inductance of an iron-core inductor with an air gap may be expressed as

$$L = \frac{0.4\pi N^2 A_c \times 10^{-8}}{l_g + l_m/\mu_r} \quad [\text{H}] \tag{8.4}$$

Inductance is seen to be inversely dependent on the effective length of the magnetic path, which is the sum of the air gap length l_g and the ratio of the core mean length to relative permeability l_m/μ_r.

When the core air gap l_g is large compared to the ratio l_m/μ_r because of the high relative permeability μ_r, variations in μ_r do not substantially affect the total effective magnetic path length or the inductance. The inductance equation then reduces to

$$L = \frac{0.4\pi N^2 A_c \times 10^{-8}}{l_g} \quad [\text{H}] \tag{8.5}$$

Final determination of the air gap requires consideration of the effect of fringing flux, which is a function of gap dimension, the shape of the pole faces, and the shape, size, and location of the winding. Its net effect is to make the effective air gap less than its physical dimension.

Fringing flux decreases the total reluctance of the magnetic path and therefore increases the inductance by a factor F to a value greater than that calculated from Equation (8.5). Fringing flux is a larger percentage of the total for larger gaps. The fringing flux factor is

$$F = 1 + \frac{l_g}{\sqrt{A_c}} \ln \left(\frac{2G}{l_g} \right) \tag{8.6}$$

where G is a dimension defined in Chapter 2. [Equation (8.6) is also valid for laminations; this equation is plotted in Figure 4.3.]

Inductance L computed in Equation (8.5) does not include the effect of fringing flux. The value of inductance L' corrected for fringing flux is

$$L' = \frac{0.4\pi N^2 A_c F \times 10^{-8}}{l_g} \quad [\text{H}] \tag{8.7}$$

The losses in an ac inductor are made up of three components: (1) Copper loss, P_{cu}; (2) iron loss, P_{fe}; and (3) gap loss, P_g. The copper loss and iron loss have been previously discussed. Gap loss is independent of core strip thickness and permeability. Maximum efficiency is reached in an inductor, as in a transformer, when the copper loss P_{cu} and the iron loss P_{fe} are equal, but only when the core gap is zero. The loss does not occur in the air gap itself but is caused by magnetic flux fringing around the gap and reentering the core in a direction of high loss. As the air gap increases, the flux across it fringes more and more, and some of the fringing flux strikes the core perpendicular to the laminations and sets up eddy currents, which cause additional loss. Distribution of fringing flux is also affected by other aspects of core geometry, the proximity of coil turns to the core, and whether there are turns on both legs (see Table 8.1). Accurate prediction of gap loss depends on the amount of fringing flux.

$$P_g = K_i E l_g f B_m^2 \quad [\text{W}] \tag{8.8}$$

where E is the strip or tongue width, cm. See Figure 3.8 and Table 3.5.

Table 8.1. Gap loss coefficient

Configuration	K_i
Two-coil C core	0.0388
Single-coil C core	0.0775
Lamination	0.1550

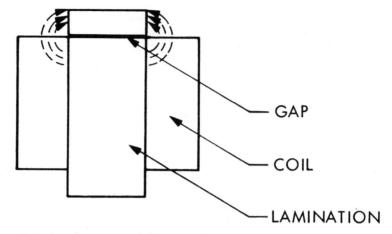

Figure 8.1. Fringing flux around the gap of an inductor designed with lamination.

The fringing flux passes around the gap and reenters the core in a direction of high loss as shown in Figure 8.1.

8.4 DESIGN EXAMPLE

For a typical design example, assume:

1. Constructed with laminations
2. Applied voltage, 115 V
3. Frequency, 60 Hz
4. Alternating current, 0.5 A
5. 25°C rise

The design procedure would then be as follows:

Step 1. Calculate the apparent power P_t from Equation (8.1):

$$P_t = VA \quad [W]$$

$$P_t = (115)(0.5)$$

$$P_t = 57.5 \quad [W]$$

Step 2. Calculate the area product A_p from Equation (8.2):

$$A_p = \left(\frac{VA \times 10^4}{4.44 B_m f K_u K_j} \right)^{1.14}$$

$B_m = 1.2$ T
$K_u = 0.4$ (see Chapter 6)
$K_j = 366$ (see Chapter 2)

$$A_p = \left(\frac{57.5 \times 10^4}{(4.44)(1.2)(60)(0.4)(366)} \right)^{1.14}$$

$$A_p = 17.4$$

Step 3. Select a size of lamination from Table 2.4 with a value A_p closest to the one calculated.

$$E1\text{-}87, \, A_p = 16.5$$

Step 4. Calculate the number of turns using Faraday's law, Equation (8.3):

$$N = \frac{E \times 10^4}{4.44 B_m f A_c}$$

The iron cross section A_c is found in Table 2.4:

$$A_c = 4.45 \quad [\text{cm}^2]$$

$$N = \frac{115 \times 10^4}{(4.44)(1.2)(60)(4.45)}$$

$$N = 808 \quad [\text{turns}]$$

Step 5. Calculate the impedance:

$$X_L = \frac{E}{I} \quad [\Omega]$$

$$X_L = \frac{115}{0.5}$$

$$X_L = 230 \quad [\Omega]$$

Step 6. Calculate the inductance:

$$L = \frac{X_L}{2\pi f} \quad [\text{H}]$$

$$L = \frac{230}{(6.28)(60)}$$

$$L = 0.610 \quad [\text{H}]$$

Step 7. Calculate the air gap from the inductance, Equation (8.5):

$$l_g = \frac{0.4\pi N^2 A_c \times 10^{-8}}{L} \quad [\text{cm}]$$

$$l_g = \frac{(1.26)(808)^2(4.45)(10^{-8})}{0.610}$$

$$l_g = 0.060 \quad [\text{cm}]$$

Gap spacing is usually maintained by inserting kraft paper. However, this paper is only available in mil thicknesses. Since l_g has been determined in cm, it is necessary to convert as follows:

$$\text{cm} \times 393.7 = \text{mils}$$

Substituting values:

$$0.060 [\text{cm}] \times 393.7 = 23.6 \quad [\text{mils}]$$

When designing inductors using lamination, it is common to place the gapping material along the mating surface between the E and I. When this method of gapping is used, only half the material is required. In this case a 10-mil and a 2-mil thickness were used.

Step 8. Calculate the amount of fringing flux from Equation (8.6); the value for G is found in Table 2.4:

$$F = 1 + \frac{l_g}{\sqrt{A_c}} \ln\left(\frac{2G}{l_g}\right)$$

$$F = 1 + \frac{0.060}{\sqrt{4.45}} \ln\left(\frac{2(3.33)}{0.060}\right)$$

$$F = 1.13$$

After finding the fringing flux F, insert it into Equation (8.7), rearrange, and solve for the correct number of turns:

$$N = \left(\frac{l_g L}{0.4\pi A_c F \times 10^{-8}}\right)^{1/2} \quad \text{[turns]}$$

$$N = \left(\frac{(0.060)(0.610)}{(1.26)(4.45)(1.13) \times 10^{-8}}\right)^{1/2}$$

$$N = 760 \quad \text{[turns]}$$

The design should be checked to verify that the reduction in turns does not cause saturation of the core.

Step 9. Calculate the current density using Table 2.1:

$$J = K_j A_p^{-0.12} \quad \text{[A/cm}^2\text{]}$$

$$J = (366)(16.5)^{-0.12}$$

$$J = 261 \quad \text{[A/cm}^2\text{]}$$

Step 10.   Determine the bare wire size $A_{w(B)}$:

$$A_{w(B)} = \frac{I}{J} \quad [\text{cm}^2]$$

$$A_{w(B)} = \frac{0.5}{261}$$

$$A_{w(B)} = 0.00192 \quad [\text{cm}^2]$$

Step 11.   Select an AWG wire size from Table 6.1, column A.

$$\text{AWG No. 24} = 0.00205 \quad [\text{cm}^2]$$

The rule is that when the calculated wire size does not fall close to those listed in the table, the next smaller size should be selected.

Step 12.   Calculate the resistance of the winding using Table 6.1, column C, and Table 2.4 for the MLT:

$$R = \text{MLT} \times N \times (\text{column C}) \times \zeta \times 10^{-6} \quad [\Omega]$$

$$R = (12.3)(760)(842.1)(1.098) \times 10^{-6}$$

$$R = 8.64 \quad [\Omega]$$

Step 13.   Calculate the power loss in the winding:

$$P_{cu} = I^2 R$$

$$P_{cu} = (0.5)^2 (8.64)$$

$$P_{cu} = 2.16 \quad [\text{W}]$$

From the core loss curves (Figure 7.10), 12-mil silicon at a flux density of 1.2 T has a core loss of approximately 1.0 mW/g. Lamination E1-87 has a weight of 481 g:

$$P_{\mathrm{fe}} = (0.001)(481)$$

$$P_{\mathrm{fe}} = 0.481 \quad [\mathrm{W}]$$

Step 14. Calculate the gap loss from Equation (8.8); the value of E is found in Table 3.5:

$$P_g = K_i E l_g f B_m^2 \quad [\mathrm{W}]$$

$$P_g = (0.155)(2.22)(0.060)(60)(1.2)^2$$

$$P_g = 1.78 \quad [\mathrm{W}]$$

Step 15. Calculate the combined losses—copper, iron, and gap:

$$P_\Sigma = P_{\mathrm{cu}} + P_{\mathrm{fe}} + P_g \quad [\mathrm{W}]$$

$$P_\Sigma = 2.16 + 0.481 + 1.78$$

$$P_\Sigma = 4.42 \quad [\mathrm{W}]$$

In a test sample made to verify these example calculations, the measured inductance was found to be 0.592 H with a current 0.515 A at 115 V, 60 Hz, and the inductor had a coil resistance of 8.08 Ω.

REFERENCE

1. Ruben, L., and Stephens, D. Gap Loss in Current-Limiting Transformers. *Electromechanical Design,* April 1973, pp. 24–126.

Chapter 9

Constant-Voltage Transformer Design

9.1 INTRODUCTION

The constant-voltage transformer has wide application, particularly where reliability and inherent regulating ability against line-voltage change are of prime importance. The output of a constant-voltage transformer is essentially a square wave, which is desirable for rectified output applications while also having good circuit characteristics. The main disadvantage to a constant-voltage transformer is efficiency and regulation for changes in frequency and load.

9.2 CONSTANT-VOLTAGE TRANSFORMER REGULATING CHARACTERISTICS

Figure 9.1 shows the basic two-coil (CVT) ferroresonant regulator, where a linear inductor L and a capacitor C are in series across an ac line. The voltage drop across the capacitor would be considerably greater than the line voltage. The voltage V_p can be limited to a predetermined amplitude by using a self-saturating transformer, which has high impedance until a certain level of flux density is reached. At that flux density, the transformer saturates and becomes a low-impedance path, which prevents further voltage buildup across the capacitor. This limiting action produces a voltage wave form that has a fairly flat top characteristic as shown in Figure 9.2 on each half-cycle.

9.3 ELECTRICAL PARAMETERS OF A CVT LINE REGULATOR

When the constant-voltage transformer is operating as a line regulator the output voltage varies as a function of input voltage, as shown in Figure 9.3. If the output of the line regulator is subjected to

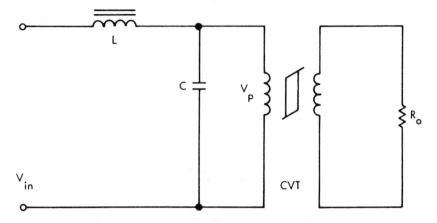

Figure 9.1. Basic circuit for a CVT.

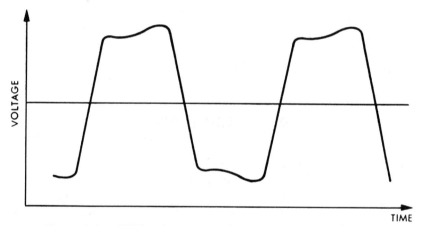

Figure 9.2. CVT voltage wave form across the primary.

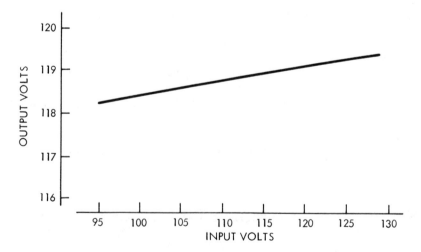

Figure 9.3. Output voltage variation as a function of input voltage.

a load power factor (lagging) less than unity, the output voltage level will be changed as shown in Figure 9.4.

If the constant-voltage transformer is subjected to a line voltage frequency change, the output will vary as shown in Figure 9.5.

The regulation of a constant-voltage transformer can be designed to be better than few percent. Capability for handling short circuits is an inherent feature of the constant-voltage transformer. The short-circuit current is limited by the series inductance L. Figure 9.6 shows the regulation characteristics at various line voltages and loads. Note that a dead short corresponding to zero output voltage does not greatly increase the load current, whereas for most transformers this would be destructive.

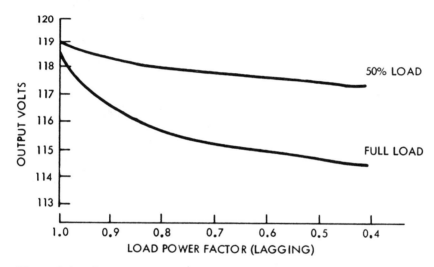

Figure 9.4. Output voltage variation as a function of load power factor.

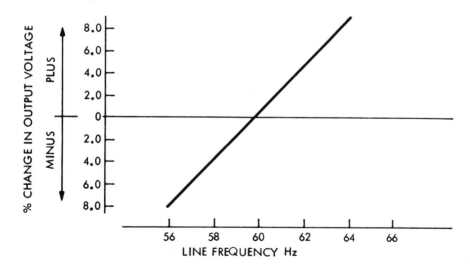

Figure 9.5. Output voltage variation as a function of line frequency change.

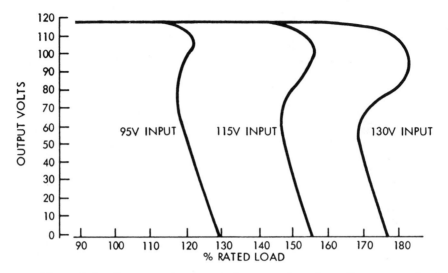

Figure 9.6. Output voltage variation as a function of output vs. load.

9.4 CONSTANT-VOLTAGE TRANSFORMER DESIGN PROCEDURES

Proper operation and power capacity of a constant-voltage transformer (CVT) depend on components L and C as shown in Figure 9.7. Experience has shown that the LC relationship is

$$LC\omega^2 = 1.5 \tag{9.1}$$

The inductance can be expressed as

$$L = \frac{\bar{R}_o}{2\omega} \quad [\text{H}] \tag{9.2}$$

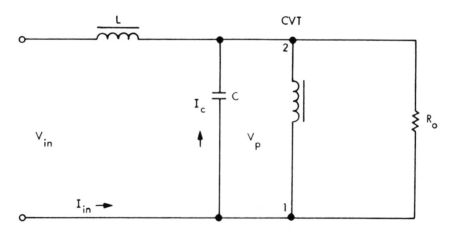

Figure 9.7. Typical CVT regulator.

and the capacitance can be expressed as

$$C = \frac{1}{0.33\omega\bar{R}_o} \quad [\text{F}] \tag{9.3}$$

Referring to Figure 9.7, assume a sinusoidal input voltage and an ideal inductor and capacitor. All voltage and currents are rms values. V_{in} is the voltage value just before the circuit starts to regulate at full load; $\bar{R}_o =$ is the reflected resistance back to the primary including efficiency,

$$\bar{R}_o = \frac{(V_p)^2\eta}{W} \quad [\Omega] \tag{9.4}$$

f is source frequency, and W is the output wattage, $W = V_s^2/R_o$.

It is common practice for the output to be isolated from the input and to connect C to a stepup winding of the constant-voltage transformer (CVT) in order to use smaller capacitor values, as shown in Figure 9.8.

The secondary current I_s can be expressed as:

$$I_s = \frac{W}{V_s} \quad [\text{A}] \tag{9.5}$$

With the stepup winding, the primary current I_p is related to the secondary current by the following equation [Ref. 3]:

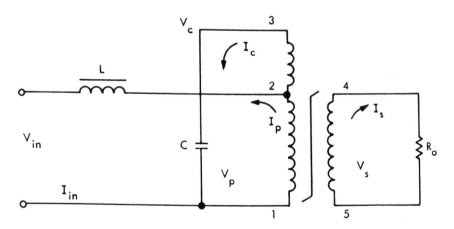

Figure 9.8. CVT with a capacitor stepup winding.

$$I_p = \frac{I_s V_{4-5}}{\eta V_{1-2}} \left(1 + \sqrt{\frac{V_{12}}{V_{13}}} \right) \quad [\text{A}] \tag{9.6}$$

The current I_c through the capacitor is increased by K because of the effective higher frequency. Due to the quasi-voltage wave form, the equivalent ac impedance of the resonant capacitor is reduced to some value lower than its normal sine wave value.

$$I_c = K V_c \omega C \quad [\text{A}] \tag{9.7}$$

where K can vary from 1.0 to 1.5.

Empirically it has been shown that for good performance the primary operating voltage should be

$$V_p = V_{\text{in (low)}} \times 0.95 \tag{9.8}$$

When the resonating capacitor is connected across a stepup winding as in Figure 9.8, where C_n is the new capacitance value and V_n is the new voltage across the capacitor,

$$C_n V_n^2 = C V_c^2 \tag{9.9}$$

The transformer power-handling capability is related to its area product by an equation* that may be stated as

$$A_p = \left(\frac{P_t \times 10^4}{4.44 B_m f K_u K_j} \right)^{1.14} \quad [\text{cm}^4] \tag{9.10}$$

The output regulation of a constant-voltage transformer for a change in line voltage is a function of the *B-H* loop characteristic as shown in Figure 9.9.

The line regulation of a constant-voltage transformer is

$$\Delta V = 4.44 \Delta B_s A_c f N \times 10^4 \tag{9.11}$$

---

*See Chapter 2 and Table 2.1.

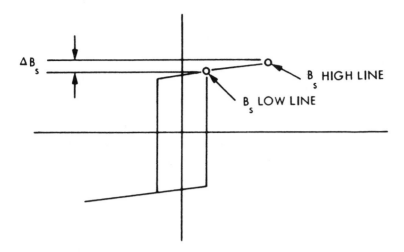

Figure 9.9. Saturated *B-H* loop at two line voltages.

9.5 CONSTANT-VOLTAGE TRANSFORMER REGULATOR DESIGN: AN EXAMPLE

Design a constant-voltage transformer regulator having an output power of 100 W at 120 V output, with an input voltage range of 105–129 V, 60 Hz. Use the circuit shown in Figure 9.8. In this design V_c is equal to 550 V for a capacitor ac rating of 660 V, a transformer efficiency of 85%, and a temperature rise of 50°C.

Step 1. Calculate the reflected resistance back to the primary including efficiency using Equation (9.4):

$$\bar{R}_o = \frac{(V_p)^2 \eta}{W} \quad [\Omega]$$

$$\bar{R}_o = \frac{(100)^2(0.85)}{100}$$

$$\bar{R}_o = 85 \quad [\Omega]$$

Step 2. Calculate the required capacitance using Equation (9.3):

$$C = \frac{1}{0.33 \omega \bar{R}_o}$$

$$C = \frac{1}{(0.33)(377)(85)}$$

$$C = 94.5 \times 10^{-6} \quad [\text{F}]$$

Step 3. Calculate the new capacitance value with the new voltage using Equation (9.9):

$$C_{13} = \frac{C_{12} V_{12}^2}{V_{13}^2}$$

$$C_{13} = \frac{(94.5)(100)^2}{(550)^2} \times 10^{-6}$$

$$C_{13} = 3.1 \times 10^{-6} \quad [\text{F}]$$

Using the standard capacitor value 3.0 μF, solve for a new value for V_{13}:

$$V_{13} = \left(\frac{C_{12} V_{12}^2}{C_{13}} \right)^{1/2} \quad [\text{V}]$$

$$V_{13} = \left(\frac{(94.5)(100)^2}{3} \right)^{1/2}$$

$$V_{13} = 560 \quad [\text{V}]$$

Step 4. Calculate the capacitance current using Equation (9.7):

$$I_c = 1.5 V_c \omega C \quad [\text{A}]$$

$$I_c = (1.5)(560)(377)(3) \times 10^{-6}$$

$$I_c = 0.950 \quad [\text{A}]$$

Step 5. Calculate the secondary current using Equation (9.5):

$$I_s = \frac{W}{V_s} \quad [\text{A}]$$

$$I_s = \frac{100}{120}$$

$$I_s = 0.833 \quad [\text{A}]$$

Step 6. Calculate the primary current using Equation (9.6):

$$I_p = \frac{I_s V_{4\text{-}5}}{\eta V_{1\text{-}2}} \left(1 + \sqrt{\frac{V_{12}}{V_{13}}} \right) \quad [\text{A}]$$

$$I_p = \frac{(0.833)(120)}{(0.85)(100)} \left(1 + \sqrt{\frac{100}{560}} \right)$$

$$I_p = 1.67 \quad [\text{A}]$$

Step 7. Calculate the apparent power P_t or the *VA* of each winding of the constant-voltage transformer:

$$P_t = VA_{12} + VA_{23} + VA_{45} \quad [\text{W}]$$

$$VA_{12} = I_p V_p \quad [\text{W}]$$

$$VA_{12} = (1.67)(100)$$

$$VA_{12} = 167 \quad [\text{W}]$$

$$VA_{23} = I_c(V_c - V_p) \quad [\text{W}]$$

$$VA_{23} = (0.950)(560 - 100)$$

$$VA_{23} = 437 \quad [\text{W}]$$

$$VA_{45} = I_s V_s \quad [W]$$

$$VA_{45} = (0.833)(120)$$

$$VA_{45} = 100 \quad [W]$$

$$P_t = 167 + 437 + 100$$

$$P_t = 704 \quad [W]$$

Step 8. Calculate the area product A_p using Equation (9.10):

$$A_p = \left(\frac{P_t \times 10^4}{4.44 B_m f K_u K_j} \right)^{1.14} \quad [cm^4]$$

$B_m = B_s = 1.95 \quad [T]$
$f = 60 \quad [Hz]$
$K_u = 0.4$ (Chapter 6)
$K_j = 534$ (Chapter 2)

$$A_p = \left(\frac{704 \times 10^4}{(4.44)(1.95)(60)(0.4)(534)} \right)^{1.14}$$

$$A_p = 113 \quad [cm^4]$$

Step 9. Select a lamination from Table 2.4 with a value of A_p closest to the one calculated:

$$E1\text{-}138, A_p = 107 \quad [cm^4]$$

Step 10. Select the core weight from Table 2.4, column 14. Then approximate the core loss in milliwatts per gram using Figure 7.11.

$$E1\text{-}138, W_t = 1880 \quad [g]$$

$$\text{Core loss at 2.0 T} \cong 3 \quad [mW/g]$$

$$(1880)(0.003) = 5.64 \quad [W]$$

Step 11. Calculate the number of primary turns using Faraday's law:

$$N_p = \frac{V_p \times 10^4}{4.44 B_m A_c f} \quad [\text{turns}]$$

$$N_p = \frac{100 \times 10^4}{(4.44)(1.95)(11.6)(60)}$$

$$N_p = 166 \quad [\text{turns}]$$

Step 12. Calculate the current density J from the data in Table 2.1:

$$J = K_j A_p^{-0.12}$$

$$J = (534)(107)^{-0.12}$$

$$J = 305 \quad [\text{A/cm}^2]$$

Step 13. Calculate the primary wire area:

$$A_{w(B)} = \frac{I_p}{J}$$

$$A_{w(B)} = \frac{1.67}{305}$$

$$A_{w(B)} = 0.00547 \quad [\text{cm}^2]$$

Step 14. Select the wire area $A_{w(B)}$ in Table 6.1 for equivalent (AWG) wire size, column A:

$$\text{AWG No. 20} = 0.00511 \quad [\text{cm}^2]$$

The rule is that when the calculated wire size does not fall close to those listed in the table, the next smaller size should be selected.

Step 15. Calculate the resistance of the primary winding, using Table 6.1, column C, and Table 2.4, column 4, for the MLT:

$$R_p = \text{MLT} \times N \times (\text{column C}) \times \zeta \times 10^{-6} \quad [\Omega]$$

$$R_p = (19.5)(166)(332)(1.20) \times 10^{-6}$$

$$R_p = 1.29 \quad [\Omega]$$

Step 16. Calculate the primary copper loss P_{cu}:

$$P_{cu} = I_p^2 R_p \quad [\text{W}]$$

$$P_{cu} = (1.67)^2(1.29)$$

$$P_{cu} = 3.60 \quad [\text{W}]$$

Step 17. Calculate the capacitor over winding number of turns:

$$N_c = \frac{N_p}{V_p} V_c - N_p$$

$$N_c = \frac{166}{100} 560 - (166)$$

$$N_c = 764 \quad [\text{turns}]$$

Step 18. Calculate the capacitor over winding wire area:

$$A_{w(B)} = \frac{I_c}{J} \quad [\text{cm}^2]$$

$$A_{w(B)} = \frac{0.95}{305}$$

$$A_{w(B)} = 0.00311 \quad [cm^2]$$

Step 19. Select the bare wire area $A_{w(B)}$ in Table 6.1 for equivalent (AWG) wire size, column A:

$$AWG \text{ No. } 22 = 0.00324 \quad [cm^2]$$

The rule is that when the calculated wire size does not fall close to those listed in the table, the next smaller size should be selected.

Step 20. Calculate the resistance of the capacitor over winding using Table 6.1, column C, and Table 2.4, column 4, for the MLT:

$$R_c = MLT \times N \times (\text{column C}) \times \zeta \times 10^{-6} \quad [\Omega]$$

$$R_c = (19.5)(764)(531)(1.20) \times 10^{-6}$$

$$R_c = 9.49 \quad [\Omega]$$

Step 21. Calculate the capacitor over winding copper loss P_{cu}:

$$P_{cu} = I_c^2 R_c \quad [W]$$

$$P_{cu} = (0.95)^2(9.49)$$

$$P_{cu} = 8.56 \quad [W]$$

Step 22. Calculate the number of secondary turns:

$$N_s = \frac{N_p}{V_p} V_s$$

$$N_s = \frac{166}{100} 120$$

$$N_s = 200 \quad \text{[turns]}$$

Step 23. Calculate the secondary winding wire area:

$$A_{w(B)} = \frac{I_s}{J} \quad \text{[cm}^2\text{]}$$

$$A_{w(B)} = \frac{0.833}{305}$$

$$A_{w(B)} = 0.00273 \quad \text{[cm}^2\text{]}$$

Step 24. Select the bare wire area $A_{w(B)}$ in Table 6.1 for equivalent (AWG) wire size, column A:

$$\text{AWG No. 23} = 0.002588 \quad \text{[cm}^2\text{]}$$

The rule is that when the calculated wire size does not fall close to those listed in the table, the next smaller size should be selected.

Step 25. Calculate the resistance of the secondary winding, using Table 6.1, column C, and Table 2.4, column 4, for the MLT:

$$R_s = \text{MLT} \times N \times (\text{column C}) \times \zeta \times 10^{-6} \quad \text{[}\Omega\text{]}$$

$$R_s = (19.5)(200)(666)(1.2) \times 10^{-6}$$

$$R_s = 3.11 \quad \text{[}\Omega\text{]}$$

Step 26. Calculate the secondary copper loss P_{cu}:

$$P_{cu} = I_s^2 R_s$$

$$P_{cu} = (0.833)^2(3.11)$$

$$P_{cu} = 2.16 \quad [W]$$

Step 27. Summarize the losses, and compare with the total losses P_Σ:

$$\text{Primary } P_{cu} = \quad 3.60$$

$$\text{Over winding } P_{cu} = \quad 8.56$$

$$\text{Secondary } P_{cu} = \quad 2.16$$

$$\text{Core } P_{fe} = \quad \underline{5.64}$$

$$\text{Total} = 20 \quad [W]$$

Step 28. Calculate the required inductance using Equation (9.2):

$$L = \frac{\bar{R}_o}{2\omega} \quad [H]$$

where $\bar{R}_o$ is the equivalent resistance at the primary with losses (see step 1), so

$$L = \frac{85}{(2)(377)}$$

$$L = 0.113 \quad [H]$$

Step 29. Calculate the short-circuit current at high line:

$$I_L = \frac{V_{in}}{X_L} \quad [A]$$

$$I_L = \frac{129}{42.6}$$

$$I_L = 3.0 \quad [\text{A}]$$

Step 30. Calculate the apparent power P_t or the *VA* of the input inductor. Using high-line voltage 129 V and the normal running current from step 6.

$$VA = V_{\text{in}} I_L \quad [\text{W}]$$

$$VA = (129)(1.67)$$

$$VA = 215 \quad [\text{W}]$$

Step 31. Calculate the area product using Equation (8.2).

$$A_p = \left(\frac{VA \times 10^4}{4.44 B_m f K_u K_j} \right)^{1.14} \quad [\text{cm}^4]$$

$B_m = 1.4$ T
$K_u = 0.4$ (Chapter 6)
$K_j = 534$ (Chapter 2)

$$A_p = \left(\frac{215 \times 10^4}{(4.44)(1.4)(60)(0.4)(534)} \right)^{1.14}$$

$$A_p = 42.8 \quad [\text{cm}^4]$$

Step 32. Select a lamination from Table 2.4 with a value of A_p closest to the one calculated:

$$\text{E1-112}, A_p = 44.9 \quad [\text{cm}^4]$$

Step 33. Calculate the number of turns using Faraday's law, Equation (8.3):

$$N = \frac{V_{\text{in}} \times 10^4}{4.44 B_m f A_c}$$

Find the iron cross section A_c in Table 2.4:

$$A_c = 5.81 \quad [\text{cm}^2]$$

Then

$$N = \frac{129 \times 10^4}{(4.44)(1.4)(60)(7.34)}$$

$$N = 471 \quad [\text{turns}]$$

Step 34. Calculate the current density J from the data in Table 2.1:

$$J = K_j A_p^{-0.12} \quad [\text{A/cm}^2]$$

$$J = (534)(44.9)^{-0.12}$$

$$J = 338 \quad [\text{A/cm}^2]$$

Step 35. Calculate the wire area for the input inductor:

$$A_{w(B)} = \frac{I_L}{J} \quad [\text{cm}^2]$$

$$A_{w(B)} = \frac{1.67}{338}$$

$$A_{w(B)} = 0.00494 \quad [\text{cm}^2]$$

Step 36. Select the bare wire area $A_{w(B)}$ in Table 6.1 for equivalent (AWG) wire size, column A:

$$\text{AWG No. } 20 = 0.005188 \quad [\text{cm}^2]$$

The rule is that when the calculated wire size does not fall close to those listed in the table, the next smaller size should be selected.

Step 37. Calculate the air gap from the inductance, Equation (8.5):

$$l_g = \frac{0.4\pi N^2 A_c \times 10^{-8}}{L} \quad [cm]$$

$$l_g = \frac{(1.26)(471)^2(7.34) \times 10^{-8}}{0.113}$$

$$l_g = 0.181 \quad [cm]$$

Gap spacing is usually maintained by inserting kraft paper. However, this paper is available only in mil thicknesses. Since l_g has been determined in cm, it is necessary to convert as follows:

$$cm \times 393.7 = mils$$

Substituting values:

$$0.181 \ [cm] \times 393.7 = 71.5 \quad [mils]$$

When designing inductors using lamination, it is common to place the gapping material along the mating surface between the E and I. When this method of gapping is used, only half of the material is required. In this case a 20-mil and a 15-mil paper were used.

$$2 \times 0.035 \times 2.54 = 0.178 \quad [cm]$$

Step 38. Calculate the amount of fringing flux from Equation (8.6); the value for G is found in Table 2.4.

$$F = 1 + \frac{l_g}{\sqrt{A_c}} \ln\left(\frac{2G}{l_g}\right)$$

$$F = 1 + \frac{0.178}{\sqrt{7.34}} \ln\left(\frac{2(4.28)}{0.178}\right)$$

$$F = 1.25$$

After finding the fringing flux F, insert it into Equation (8.7), rearrange, and solve for the correct number of turns:

$$N = \left(\frac{l_g L}{0.4\pi A_c F \times 10^{-8}}\right)^{1/2} \quad \text{[turns]}$$

$$N = \left(\frac{(0.178)(0.113)}{(1.26)(7.34)(1.25) \times 10^{-8}}\right)^{1/2}$$

$$N = 417 \quad \text{[turns]}$$

The design should be checked to verify that the reduction in turns does not cause saturation of the core.

$$B_m = \frac{V \times 10^4}{4.44 N A_c f}$$

$$B_m = \frac{129 \times 10^4}{(4.44)(417)(7.34)(60)}$$

$$B_m = 1.58 \quad \text{[T]}$$

From the core loss curves (7.11), 12-mil silicon at a flux density of 1.58 T has a core loss of approximately 2.0 mW/g. Lamination E1-120 has a weight of 1029 g (see Table 2.4).

$$P_{\text{fe}} = (0.002)(1029)$$

$$P_{\text{fe}} = 2.06 \quad \text{[W]}$$

Step 39. Calculate the gap loss from Equation (8.8); the value of D is found in Table 2.4.

$$P_g = K_i D l_g f B_m^2 \quad [\text{W}]$$

$$P_g = (0.155)(2.54)(0.178)(60)(1.4)^2$$

$$P_g = 8.24 \quad [\text{W}]$$

Step 40. Calculate the resistance of the inductor winding using Table 6.1, column C, and Table 2.4, column 4, for the MLT.

$$R_L = \text{MLT} \times N \times (\text{column C}) \times \zeta \times 10^{-6} \quad [\Omega]$$

$$R_L = (16.0)(417)(332)(1.2) \times 10^{-6}$$

$$R_L = 2.66 \quad [\Omega]$$

Step 41. Calculate the inductor winding copper loss P_{cu}:

$$P_{\text{cu}} = I_L^2 R_L \quad [\text{W}]$$

$$P_{\text{cu}} = (1.67)^2(2.66)$$

$$P_{\text{cu}} = 7.42 \quad [\text{W}]$$

Step 42. Calculate the combined losses—copper, iron, and gap:

$$P_\Sigma = P_{\text{cu}} + P_{\text{fe}} + P_g \quad [\text{W}]$$

$$P_\Sigma = 7.42 + 2.06 + 8.24$$

$$P_\Sigma = 17.5 \quad [\text{W}]$$

The magnetic components were wound to verify these example calculations. The test results are shown in Table 9.1. The constant-voltage transformer was built, and the electrical data recorded in Table 9.2 and plotted in Figure 9.10.

Table 9.1.

Inductance		Resistance	
Calculated	Measured	Calculated	Measured x 1.2
0.113	0.135	2.66	2.60
Transformer		Resistance	
Winding		*Calculated	Measured x 1.2
1 - 2		1.29	1.77
2 - 3		9.49	9.84
4 - 5		3.11	3.78

*All winding were calculated with the same MLT.

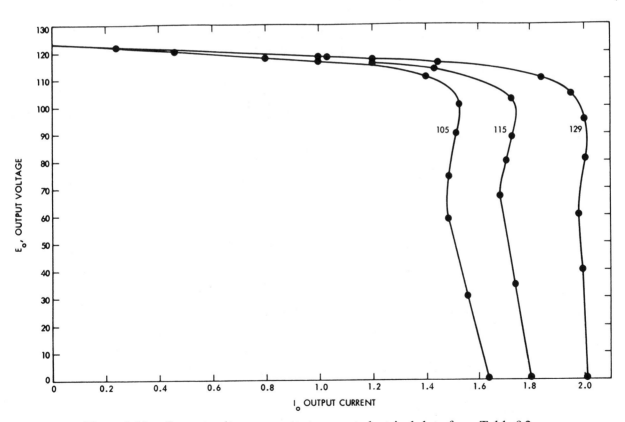

Figure 9.10. Output voltage vs. output current electrical data from Table 9.2.

Table 9.2. CVT input-output electrical data

R_L	115				105V				129V			
	I_{in}	I_o	E_o	I_c	I_{in}	I_o	E_o	I_c	I_{in}	I_o	E_o	I_c
0	1.575	1.76	0	0.0363	1.425	1.6	0	0.0332	1.75	2.1	0	0.0412
20	1.675	1.72	35	0.205	1.51	1.56	31.6	0.184	1.925	2.0	40.5	0.237
40	2.0	1.68	67.6	0.363	1.77	1.48	59.3	0.319	2.32	2.1	81.2	0.438
60	2.11	1.73	104	0.550	2.04	1.51	90.5	0.477	1.97	1.86	112	0.620
80	1.55	1.45	116	0.672	1.72	1.41	113	0.624	1.45	1.46	117	0.739
100	1.25	1.20	118	0.741	1.31	1.19	117	0.695	1.22	1.20	118	0.821
120	1.04	1.00	119	0.798	1.075	0.992	118	0.746	1.05	1.00	120	0.878
150	0.86	0.81	120	0.847	0.85	0.80	119	0.793	0.91	0.81	120	0.930
180	0.76	0.68	120	0.877	0.76	0.678	120	0.822	0.82	0.68	121	0.959
200	0.65	0.57	121	0.901	0.65	0.57	120.6	0.843	0.75	0.56	121	0.981
250	0.575	0.458	120	0.917	0.55	0.450	120	0.864	0.68	0.46	121	1.00
300	0.525	0.385	121	0.928	0.50	0.380	120	0.874	0.65	0.385	121	1.01
400	0.455	0.28	121	0.943	0.45	0.28	120.6	0.888	0.65	0.28	122	1.02
500	0.45	0.23	121	0.950	0.40	0.28	121	0.892	0.55	0.23	122	1.03

REFERENCES

1. H. P. Hart and R. J. Kakalec, "The derivation and application of design equations for ferroresonant voltage regulators and regulated rectifiers," *IEEE Trans. Magnetics,* vol. Mag-7, No. 1, March 1971, pp. 205–211.

2. I. B. Friedman, "The analysis and design of constant voltage regulators," *IRE Trans. Component Parts,* vol. CP-3, March 1956, pp. 11–14.

3. S. Lendena, "Design of a magnetic voltage stabilizer," *Electronics Technology,* May 1961, pp. 154–155.

Chapter 10

Current Transformer Design

10.1 INTRODUCTION

Current transformers are used to measure or monitor the current in the lead of an ac power circuit. They are very useful in high-power circuits where the currents are large, i.e., higher than the ratings of so-called self-contained current meters. Other applications relate to overcurrent and undercurrent relaying for power circuit protection, such as in the power lead of an inverter or converter. Multiturn secondaries then provide a reduced current for detecting overcurrent, undercurrent, peak current, and average current, as shown in Figure 10.1.

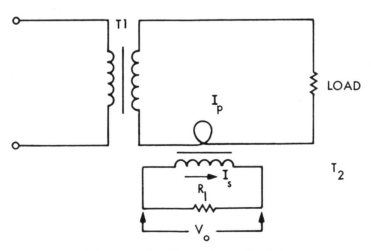

Figure 10.1. Current monitor T2.

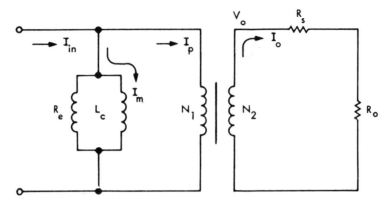

Figure 10.2. Simplified equivalent circuit for a current transformer.

In current transformer design, the core characteristics must be carefully selected because excitation current I_m essentially subtracts from the metered current and affects the ratio and phase angle of the output current. The simplified equivalent circuit of a current transformer shown in Figure 10.2 represents the important elements of a current transformer where the ratio of primary to secondary turns is

$$n = \frac{N_2}{N_1} \qquad (10.1)$$

10.2 ANALYSIS OF THE INPUT CURRENT COMPONENT

A better understanding of current transformer behavior may be achieved by considering input in the primary winding in terms of various components. Only ampere-turn component $I_{in}N_1$ drives the magnetic flux around the core. Ampere-turn I_mN_1 provides the core loss. The secondary ampere-turns, I_sN_2, balance the remainder of the primary ampere-turns.

The exciting current I_m in Figure 10.2 determines the maximum accuracy that can be achieved with a current transformer. *Exciting current* may be defined as the portion of the primary current that satisfies the hysteresis and eddy current losses of the core. If the values of L_c and R_e in Figure 10.2 are too small because the permeability of the core material is low and the core loss is high, only a part of the current (nI_p) will flow in the output load R_o. Figure 10.3 shows how the exciting current relates to the output. The exciting current is equal to

$$I_m = \frac{Hl_m}{0.4\pi N} \quad [A] \qquad (10.2)$$

where H is the magnetizing force and l_m is the core mean length.

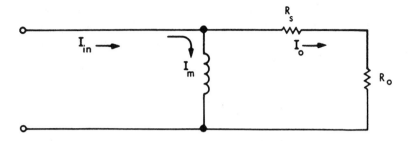

Figure 10.3. Input current–output current relationship.

The input current is made up of two components:

$$I_{in}^2 = I_m^2 + I_o^2 \tag{10.3}$$

Therefore,

$$I_m^2 = I_{in}^2 - I_o^2 \tag{10.4}$$

$$I_m = I_{in}\left[1 - \left(\frac{I_o}{I_{in}}\right)^2\right]^{1/2} \tag{10.5}$$

Figure 10.4 shows graphically that the higher the exciting current or core loss, the larger the error.

The secondary load R_o, secondary winding resistance R_s, and secondary current I_s determine the induced voltage of a current transformer.

$$V_o = I_s(R_s + R_o) \tag{10.6}$$

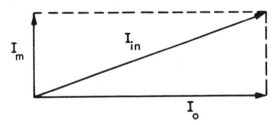

Figure 10.4. Diagram of phase relationship.

This voltage determines the number of turns. The value of B_m should be chosen as high as the level that gives maximum permeability; it can be higher if the excitation currents are well below and do not disturb the current ratio. The core cross-sectional area required to generate a voltage in the secondary circuit is

$$A_c = \frac{I_s(R_s + R_o) \times 10^4}{K N_2 f B_m} \quad [\text{cm}^2] \tag{10.7}$$

When choosing a core material, a reasonable value for B_m typically results in L_c and R_e values (Figure 10.2) large enough to reduce the current flowing in these elements so as to satisfy the ratio and phase requirements. The inductance is calculated from the equation

$$L_c = \frac{0.4 \pi N_1^2 A_c \Delta \mu}{l_m \times 10^8} \quad [\text{H}] \tag{10.8}$$

The value of the equivalent resistance R_e is obtained from

$$R_e = \frac{(I_s R_s)^2}{P_{\text{cu}}} \quad [\Omega] \tag{10.9}$$

When

$$\frac{R_e}{n^2} \gg R_s + R_o \tag{10.10}$$

and

$$\frac{2\pi f L_c}{n^2} \gg R_s + R_o \tag{10.11}$$

then

$$I_p = n I_s \tag{10.12}$$

or

$$I_p N_p = I_s N_s \tag{10.13}$$

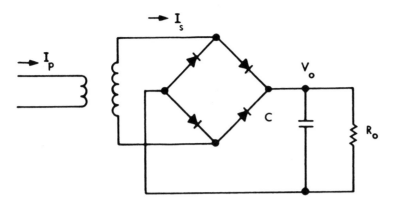

Figure 10.5. Current transformer with dc output.

Except for relatively low-accuracy industrial types, current transformers are wound on toroidal cores, which virtually eliminates errors due to leakage flux. Some errors may be compensated for by adjusting the number of secondary turns.

A current transformer with a dc output is shown in Figure 10.5.

10.3 CURRENT TRANSFORMER DESIGN: AN EXAMPLE

Assume that specifications for the design of a current transformer as shown in Figure 10.5 require the following:

1. One-turn primary
2. $I_{in} = 0-5$ A
3. $V_o = 0-5$ V
4. $R_o = 500$ $\Omega 1\%$
5. $f = 2500$ Hz (square wave)
6. $B_m = 0.2$ T
7. Core loss less then 3% (error)

Step 1. Calculate the secondary current:

$$I_s = \frac{V_o}{R_o} \quad [A]$$

$$I_s = \frac{5.0}{500}$$

$$I_s = 0.01 \quad [A]$$

Step 2. Calculate the secondary turns using Equation (10.13):

$$N_s = \frac{I_p N_p}{I_s} \quad [\text{turns}]$$

$$N_s = \frac{(5)(1)}{0.01}$$

$$N_s = 500 \quad [\text{turns}]$$

Step 3. Calculate the core iron cross section using Equation (10.7), allowing for a 1.0-V diode drop (V_d):

$$A_c = \frac{(V_o + V_d) \times 10^4}{K B_m f N} \quad [\text{cm}^2]$$

$$K = 4.0 \text{ (square wave)}$$

$$A_c = \frac{(5.0 + 2.0) \times 10^4}{(4.0)(0.2)(2500)(500)}$$

$$A_c = 0.07 \quad [\text{cm}^2]$$

Step 4. Select a toroidal core from Table 2.7 with a value A_c closest to the one calculated:*

$$\text{Core } 52000, A_c = 0.086 \quad [\text{cm}^2]$$

Step 5. Calculate the effective window area $W_{a(\text{eff})}$:

$$W_{a(\text{eff})} = W_a S_3 \quad [\text{cm}^2]$$

A typical value for S_3 is 0.75, as shown in Chapter 6.

Select the window area W_a from Table 2.7 for 52000:

---

*Cores in Table 2.7 have 2-mil tape thickness.

$$W_{a(\text{eff})} = (0.982)(0.75)$$

$$W_{a(\text{eff})} = 0.737 \quad [\text{cm}^2]$$

The secondary wire occupies half of the effective window area:

$$W_{a(\text{sec})} = 0.368 \quad [\text{cm}^2]$$

Step 6. Calculate the wire area A_w with insulation, using a fill factor S_2 of 0.6:

$$A_w = \frac{W_{a(\text{sec})}}{N} S_2 \quad [\text{cm}^2]$$

$$A_w = \frac{0.368}{500}(0.6)$$

$$A_w = 0.00044 \quad [\text{cm}^2]$$

Step 7. Select the wire area A_w with insulation in Table 6.1 for equivalent AWG wire size column D:

$$\text{AWG No. 32} = 0.000456$$

The rule is that when the calculated wire size does not fall close to those listed in the table, the next smaller size should be selected.

Step 8. Calculate the resistance of the secondary winding, using Table 6.1, column C, and Table 2.7, for the MLT:

$$R_s = \text{MLT} \times N \times (\text{column C}) \times 10^{-6} \quad [\Omega]$$

$$R_s = (2.7)(500)(5315) \times 10^{-6}$$

$$R_s = 7.0 \quad [\Omega]$$

Step 9. Calculate secondary output power:

$$P_o = (V_o + V_d)(I_s) \quad [\text{W}]$$

$$P_o = (5 + 2)(0.01)$$

$$P_o = 0.07 \quad [\text{W}]$$

Step 10. Calculate acceptable core loss:

$$P_{\text{fe}} = P_o\left(\frac{\text{core loss, \%}}{100}\right) \quad [\text{W}]$$

$$P_{\text{fe}} = 0.07\left(\frac{2}{100}\right)$$

$$P_{\text{fe}} = 0.0014 \quad [\text{W}]$$

Step 11. Select the core weight from Table 2.7, column 14; then calculate the core loss in mW/g:

$$\text{Core 52000,} \ W_t = 3.73 \quad [\text{g}]$$

$$\frac{P_{\text{fe}}}{W_t} \times 10^3 = \text{mW/g}$$

$$\frac{0.0014}{3.73} \times 10^3 = \text{mW/g}$$

$$0.375 \quad \text{mW/g}$$

Step 12. Select the proper magnetic material in Figure 3.5, reading from the 2.5-kHz frequency curve for a flux density of 0.2 T. The magnetic material that comes closest to 0.375 mW/g is

Supermalloy, with 0.5 mW/g. When nickel steel is used, Table 7.1 provides a weight correction factor.

The weight from Table 2.7 is multiplied by the weight correction factor:

$$3.73 \times 1.148 = 4.28 \quad [g]$$

With a weight of 4.28 g, the total core loss is

$$0.5 \times 4.28 \times 10^{-3} = 0.0021 \quad [W]$$

Step 13. With this new core loss, calculate the new core error in percent:

$$\text{Core loss, \% (error)} = \frac{P_{fe}}{P_o} \times 100 \quad [\%]$$

$$\text{Core loss, \% (error)} = \frac{0.0021}{0.07} \times 100$$

$$\text{Core loss, \% (error)} = 3.0 \quad [\%]$$

A current transformer was built and the data recorded in Table 10.1 and plotted in Figure 10.6 with an error of 3.4%. The secondary winding resistance was 6.5 Ω.

Table 10.1. Current transducer electrical data

I_{in}	E_o	I_{in}	E_o
0.250	0.227	2.693	2.593
0.500	0.480	3.312	3.181
0.746	0.722	3.625	3.488
1.008	0.978	3.942	3.791
1.262	1.219	4.500	4.339
1.441	1.377	5.014	4.831
2.010	1.929	5.806	5.606
2.400	2.310		

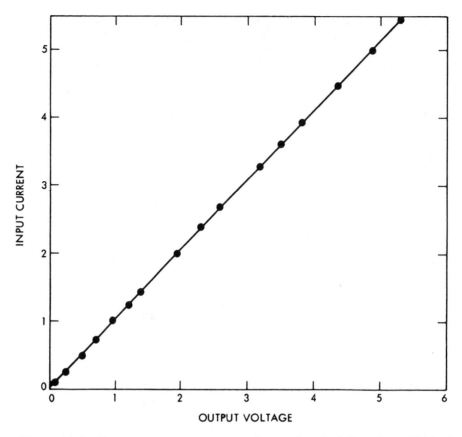

Figure 10.6.   Input current vs. output voltage; electrical data from Table 10.1.

REFERENCES

1. *Tape Wound Cores,* Bulletin TC-101B, Arnold Engineering, Marengo, Illinois, undated.

2. *Tape Wound Core Design Manual,* Bulletin C1-171-20M, Magnetic Metals, Camden, New Jersey, undated.

3. William Dull, *Designer's Guide to Current and Power Transformers,* EDN, March 5, 1975, pp. 47–52.

Chapter 11

Three-Phase Transformer Design

11.1 INTRODUCTION

Three-phase power is used almost exclusively for generation, transmission, and distribution and for all industrial uses. It has many advantages over single-phase power. Transformers can be made smaller and lighter for the same power-handling capability because the copper and iron are used more effectively. In circuitry for conversion from ac to dc, the output contains a much lower ripple amplitude and higher frequency component 3×, 6× line frequency, which requires less filtering.

Two connection arrangements are in common use for three-phase transformers, one known as a star or Y connection and the other known as a delta (Δ) connection. The design requirements for each particular job dictate which method of connection should be used.

11.2 COMPARING TRANSFORMER PHYSICAL SIZES

The schematic diagram in Figure 11.1 shows the connection of three single-phase transformers operating from a three-phase power source and a single three-phase EI lamination operating from a three-phase power source connected in a delta-delta configuration. The single three-phase transformer T4 will be lighter and small than a bank of three single-phase transformers of the same total rating. Since the windings of the three-phase transformers are placed on a common magnetic core rather than on three independent cores, the consolidation results in an appreciable saving in copper, core, and insulating materials.

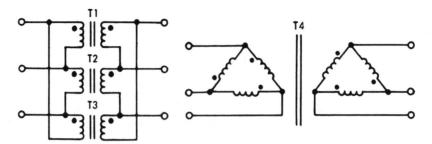

Figure 11.1. Transformer schematic.

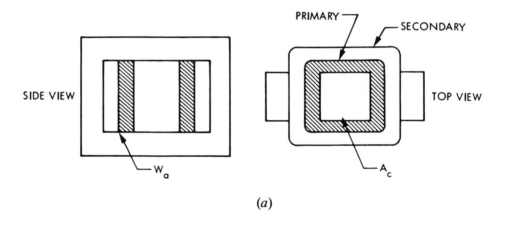

(a)

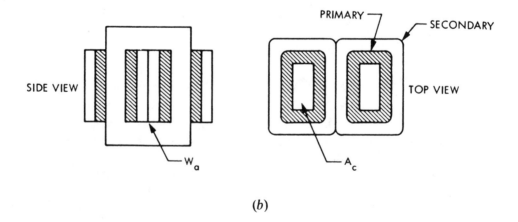

(b)

Figure 11.2. Cutaway views of a single phase transformer with (a) an *EI* lamination core; (b) a C core.

Figure 11.2 presents cutaway views of a single-phase transformer showing the window area and iron area of two types of core configurations. The *EI* lamination shown in Figure 11.2*a* is known as a *shell type* because it looks like the core surrounds the coil. The C core shown in Figure 11.2*b* is known as a *core type* because it looks like the coil surrounds the core.

Cutaway views of a three-phase transformer are shown in Figure 11.3. These cross-sectional views show the window and iron areas. The three-legged core is designed to take advantage of the fact that with balanced voltages impressed the flux in each phase leg adds up to zero. Therefore, no return leg is needed under normal conditions. When the transformer is subjected to unbalanced loads or unbalanced line voltages it may be best to use three single-phase transformers because of the high circulating currents.

11.3 PHASE AND LINE CURRENT AND VOLTAGE IN A DELTA SYSTEM

In a three-phase delta circuit such as the one shown in Figure 11.4, the line voltage and line current are commonly called *phase voltage* and *phase current*. The line voltage E_{line} will be the same as the actual winding voltage of the transformer. However, the line current I_{line} is equal to the phase current I_{phase} times $\sqrt{3}$:

$$I_{\text{line}} = I_{\text{phase}} \sqrt{3} \qquad (11.1)$$

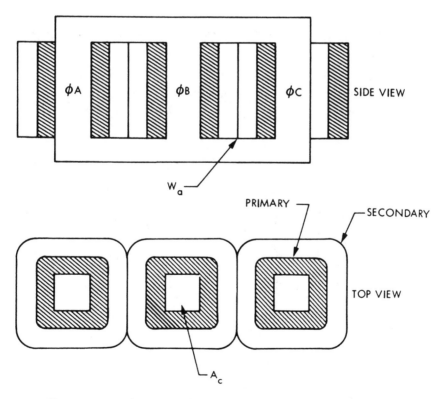

Figure 11.3. Cutaway views of a three-phase transformer.

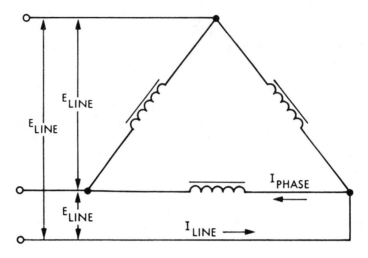

Figure 11.4. Three-wire delta.

so that

$$I_{\text{phase}} = \frac{I_{\text{line}}}{\sqrt{3}} \qquad (11.2)$$

11.4 PHASE AND LINE CURRENT AND VOLTAGE IN A WYE SYSTEM

The relationship between the line voltage and line current and the winding or phase voltage and phase current in a three-phase wye circuit can be seen in Figure 11.5. In a wye system, the voltage between any two wires in the lines will always be $\sqrt{3}$ times the phase voltage E_{phase} between the

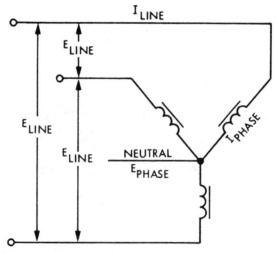

Figure 11.5. Four-wire wye.

neutral and any one of the lines:

$$E_{line} = E_{phase} \sqrt{3} \qquad (11.3)$$

$$E_{phase} = \frac{E_{line}}{\sqrt{3}} \qquad (11.4)$$

The line current in the wye connection is equal to the phase current.

11.5 THE AREA PRODUCT A_p AND ITS RELATIONSHIPS IN THREE-PHASE TRANSFORMERS

The A_p of a three-phase core (see Chapter 2) is defined differently than that for a single-phase core. Figures 11.6 and 11.7 show W_a and A_c for single-phase and three-phase transformers, respectively. The A_p of a core is the product of the available window area W_a of the core in square centimeters (cm^2) multiplied by the effective cross-sectional area A_c in square centimeters (cm^2), which may be stated as

$$\text{Single-phase:} \quad A_p = W_a A_c \quad [\text{cm}^4] \qquad (11.5)$$

This is all right for single-phase transformers. But for three-phase transformers, because there are basically two window areas and three iron areas, the window utilization is different, and the area product changes to

$$\text{Three-phase:} \quad A_p = \frac{3}{2} W_a A_c \quad [\text{cm}^4] \qquad (11.6)$$

Figure 11.8 shows the outline form of a three-phase transformer core that is typical of those shown in suppliers' catalogs.

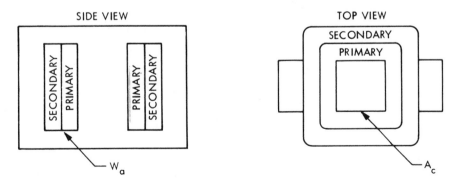

Figure 11.6. W_a and A_c in a single-phase EI transformer.

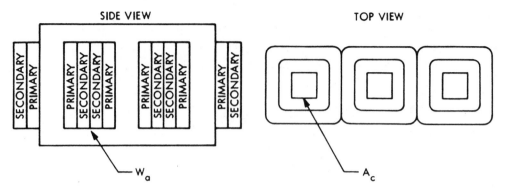

Figure 11.7. W_a and A_c in a three-phase EI transformer.

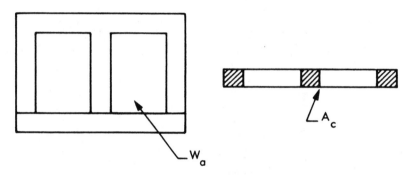

Figure 11.8.   Three-phase EI lamination.

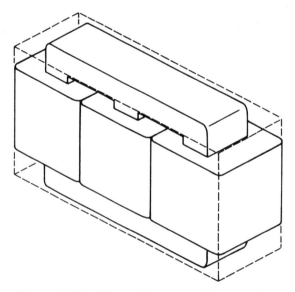

Figure 11.9.   Three-phase transformer volume.

Table 11.1. Three-phase lamination core configuration constants

CORE	LOSSES	K_i 25°C	K_i 50°C	K_s	K_w	K_v
THREE* PHASE LAMINATION	$P_{cu} = P_{fe}$	304	443	52.2	91.4	23.8
$J = K_i A_p^{-0.12}$; $A_t = K_s A_p^{0.5}$; $W_t = K_w A_p^{0.75}$; $Vol = K_v A_p^{0.75}$						

*The three-phase core is also available in double EE built from two C cores in a large variety of sizes. The manufacturers are Arnold Engineering Co., Magnetic Metals, and National Magnetics.

There is a unique relationship between the area product A_p characteristic number for transformer cores and several other important parameters that must also be consdiered in three-phase transformer design.

Table 11.1 was developed from the data obtained in Table 11.2. The area product A_p relationships with volume, surface area, current density, and weight are presented in detail in Chapter 2.

11.6 CALCULATION OF THREE-PHASE TRANSFORMER VOLUME

The volume of a transformer can be related to its area product A_p, treating the volume as shown in Figure 11.9 as a solid quantity. The volume/area product relationship is

$$\text{Volume} = K_v A_p^{0.75} \tag{11.7}$$

in which K_v is a constant related to core configuration. This constant was obtained by averaging the values in Table 11.2, column 15. The derivation for this equation is in Chapter 2.

The relationship between volume and area product A_p is illustrated in Figure 11.10. It was obtained from the data shown in Table 11.2, columns 3 and 15.

11.7 CALCULATION OF THREE-PHASE TRANSFORMER WEIGHT

The total weight W_t of a transformer can be related to the area product A_p of a transformer by the equation

$$W_t = K_w A_p^{0.75} \tag{11.8}$$

in which K_w is a constant related to core geometry. This constant was obtained by averaging the values in Table 11.2, column 14. The derivation of Equation (11.8) is given in Chapter 2.

The relationship between weight and area product is illustrated in Figure 11.11. It was obtained from the data shown in Table 11.2, columns 3 and 14.

Table 11.2. Lamination characteristics

		1	2	3		4	5		6	7	8
				A_p, cm^4			N				$I = \sqrt{\dfrac{W}{\Omega}}$
		Core	A_t, cm^2		per leg	MLT, cm		AWG	Ω @ 50°C	P_Σ	
1		EI-1/4	73.7	1.36		4.28	288		0.449	2.21	1.57
					0.454			20			
2		EI-3/8	128	4.95		6.13	484		1.08	3.84	1.33
					1.65			20			
3		EI-1/2	181	12.1		7.86	674		1.93	5.43	1.18
					4.03			20			
4		EI-9/16	252	23.7		8.42	1030		3.16	7.56	1.09
					7.89			20			
5		EI-5/8	319	38.5		9.69	1400		4.95	9.57	0.983
					12.83			20			
6		EI-7/8	556	113		13.4	2143		10.5	16.7	0.892
					37.7			20			
7		EI-1.00	844	253		16.4	2033		6.07	25.3	1.44
					84.2			17			
8		EI-1.20	817	291		17.2	1602		5.02	24.5	1.56
					97.1			17			
9		EI-1.50	1293	690		21.4	2558		9.97	38.8	1.39
					230			17			
10		EI-1.80	1839	1470		25.7	3744		17.5	55.2	1.25
					491			17			
11		EI-2.40	3270	4680		34.2	6764		42.1	98.1	1.08
					1560			17			
12		EI-3.60	7373	14100		51.1	15393		143	221	0.878
					4700			17			

Definitions for Table 11.2

Information given is listed by column as:

1. Manufacturer part number (Thomas & Skinner Inc.)
2. Surface area calculated from Figure 11.12.
3. Area product effective iron area times window area, total A_p and A_p/leg
4. Mean length turn on one bobbin
5. Total number of turns and wire size for all three bobbins using a window utilization factor $K_u = 0.40$
6. Resistance of the wire at 50°C
7. Watts loss is based on Figure 7.2 for a ΔT of 25°C with a room ambient temperature of 25°C surface dissipation times the transformer surface area; total loss is equal to 2 P_{cu}
8. Current calculated from columns 6 and 7

9	10	11	12	13	14		15	16
$\Delta T(25°C)$ $J = I/\text{cm}^2$	Ω @ 75°C	P_Σ	$I = \sqrt{\dfrac{W}{\Omega}}$	$\Delta T(50°C)$ $J = I/\text{cm}^2$	Weight		Volume, cm^3	A_c, cm^2 (14 mil)
					fe	cu		
302	0.494	5.16	2.28	440	56.5	58.5	31.4	0.363
257	1.19	8.96	1.94	374	162	140	84.6	0.817
228	2.12	12.7	1.73	333	341	250	163	1.45
210	3.47	17.6	1.59	307	443	410	234	1.84
189	5.44	22.3	1.43	276	743	641	367	2.27
172	11.5	38.9	1.30	251	1835	1357	875	4.45
139	6.66	59.1	2.10	202	2896	3144	1,650	5.805
150	5.51	57.2	2.28	219	3733	2598	1,530	8.37
134	10.9	90.5	2.03	196	7323	5135	2,990	12.7
121	19.2	129	1.83	176	12650	9074	5,160	18.8
104	46.3	229	1.57	151	30141	21799	12,200	33.6
84	157	516	1.28	123	101900	74200	41,300	44.8

 9. Current density calculated from columns 5 and 8
10. Resistance of the wire at 75°C
11. Watts loss is based on Figure 7.2 for a ΔT of 50°C with a room ambient of 25°C surface dissipation times the transformer surface area; total loss is equal to 2 P_{cu}
12. Current calculated from columns 10 and 11
13. Current density calculated from columns 5 and 12
14. Effective core weight for silicon plus copper; weight in grams
15. Transformer volume calculated from Figure 11.9
16. Core effective cross section (thickness, 0.014), square stack

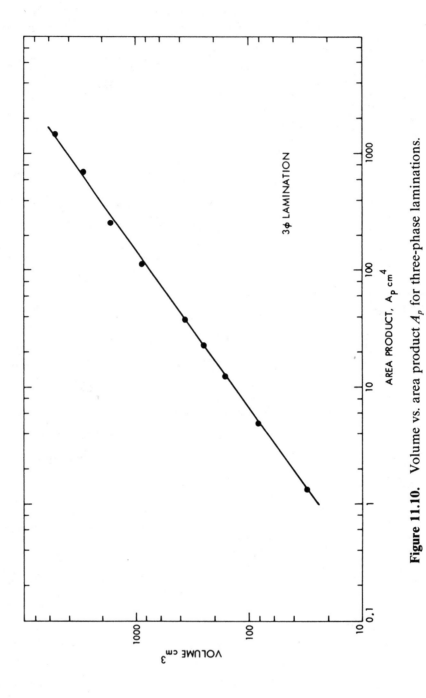

Figure 11.10. Volume vs. area product A_p for three-phase laminations.

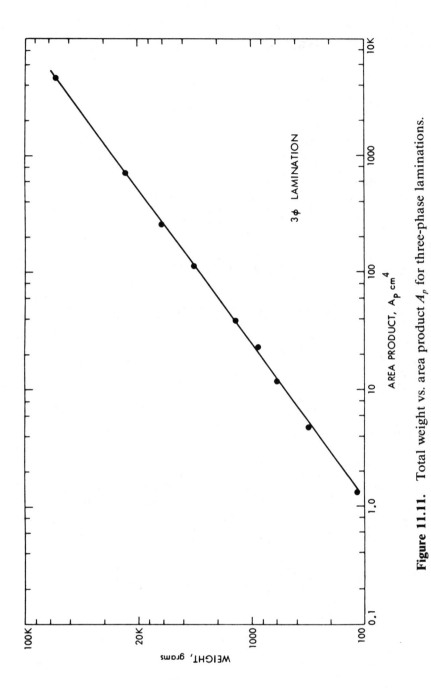

Figure 11.11. Total weight vs. area product A_p for three-phase laminations.

11.8 CALCULATION OF THREE-PHASE TRANSFORMER SURFACE AREA

The surface area A_t of a transformer can be related to the area product A_p of a transformer, treating the surface area as shown in Figure 11.12. This relationship can be expressed as

$$A_t = K_s A_p^{0.5} \qquad (11.9)$$

in which K_s is a constant related to core configuration. This constant was obtained by averaging the values in Table 11.2, column 2. The derivation of Equation (11.9) is in Chapter 2.

The relationship between surface area and area product is illustrated in Figure 11.13. It was obtained from the data shown in Table 11.2, columns 2 and 3.

11.9 CALCULATION OF THREE-PHASE TRANSFORMER CURRENT DENSITY

Current density J of a transformer can be related to its area product A_p for a given temperature rise by the equation

$$J = K_j A_p^{-0.12} \qquad (11.10)$$

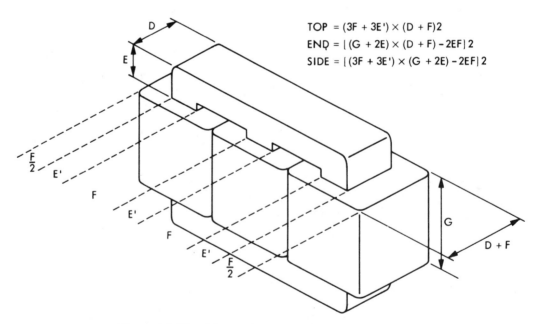

TOP = (3F + 3E') × (D + F)2
END = [(G + 2E) × (D + F) − 2EF]2
SIDE = [(3F + 3E') × (G + 2E) − 2EF]2

Figure 11.12. Three-phase transformer surface area.

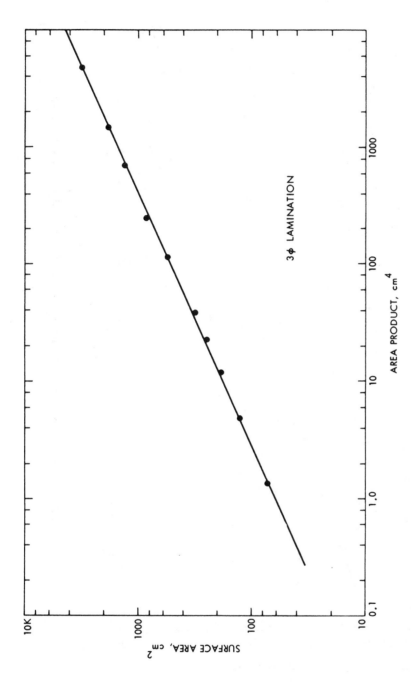

Figure 11.13. Surface area vs. area product A_p for three-phase laminations.

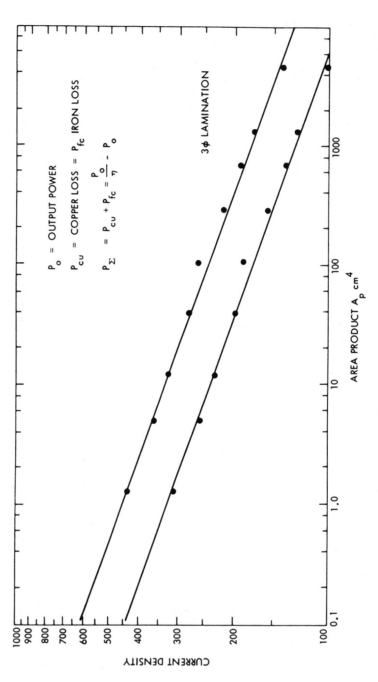

Figure 11.14. Current density vs. area product A_p for 25°C and 50°C rises for three-phase laminations.

in which K_j is a constant related to core configuration. This constant was obtained by averaging the values in Table 11.2, columns 9 and 13. The derivation of Equation (11.10) is in Chapter 2.

The relation between current density and area product is given in Figure 11.14 for temperature increases of 25°C and 50°C. It was obtained from data given in Table 11.2, columns 3, 9, and 13.

11.10 RELATIONSHIP OF A_p TO POWER-HANDLING CAPABILITY IN A THREE-PHASE TRANSFORMER

Using the approached developed in Chapter 3 for single-phase transformers, the power-handling capability of a three-phase core can be related to its area product by the equation

$$A_p = \left(\frac{P_t \times 10^4}{4.44 B_m f K_u K_j}\right)^{1.14} \quad [\text{cm}^4] \tag{11.11}$$

where B_m is the flux density, T

f is the frequency, Hz

K_u is the window utilization factor (see Chapter 6)

K_j is the current density coefficient (see Table 11.1)

P_t is the apparent power, primary plus secondary (see Chapter 3)

Apparent power P_t is described in detail in Chapter 3. The apparent power of the transformer is the combined power of primary and secondary windings, which handle P_{in} and P_o to the load, respectively. Since the power transformer has to be designed of accommodate the primary P_{in} and secondary P_o,

$$P_t = P_{in} + P_o \tag{11.12}$$

$$P_{in} = \frac{P_o}{\eta} \tag{11.13}$$

Substituting (11.13) into (11.12),

$$P_t = \frac{P_o}{\eta} + P_o \tag{11.14}$$

$$P_t = P_o\left(\frac{1}{\eta} + 1\right) \tag{11.15}$$

The designer must be concerned with the apparent power handling capability, P_t, of the transformer core and windings. P_t varies with the type of circuit in which the transformer is used. If the current in the rectifier transformer is interrupted, its effective rms value changes. Transformer size is thus determined not only by the load demand but also by current wave shape. For example, for a load of 1 W, 1 V, and 1 A, compare the power-handling capability required for each winding (neglecting transformer and diode losses so that $P_{in} = P_o$) for the three-phase half-wave delta-wye of Figure 11.15, the three-phase full-wave delta-wye of Figure 11.16, and the three-phase full-wave delta-delta of Figure 11.17. (Data are from Table 11.3.)

The total apparent power P_t for the circuit shown in Figure 11.15 is 2.69 W.

$$\text{Primary} = 1.21 \tag{11.16}$$

$$\text{Secondary} = 1.48 \tag{11.17}$$

$$P_t = P_o\left(\frac{1.21}{\eta} + 1.48\right) \quad [\text{W}] \tag{11.18}$$

$$V_1 = 0.855V_o \quad [\text{V}] \tag{11.19}$$

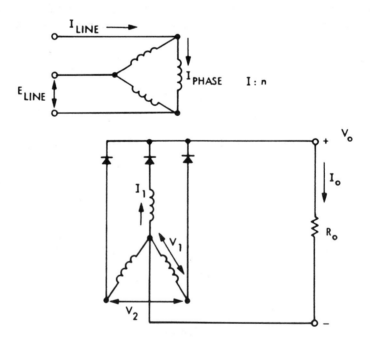

Figure 11.15. Three-phase half-wave delta-wye.

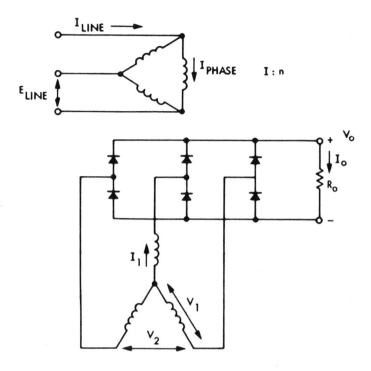

Figure 11.16. Three-phase full-wave delta-wye.

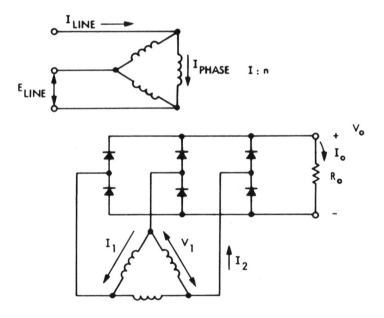

Figure 11.17. Three-phase full-wave delta-delta.

Table 11.3. Three-phase rectifier circuit data*

	delta-wye half wave	delta-wye full wave	delta-delta full wave
V_o average	1.000	1.000	1.000
I_o average	1.000	1.000	1.000
Line voltage	0.855	0.428	0.740
Line current	0.816	1.410	0.816
Primary V/leg	0.855	0.428	0.740
Primary I/leg	0.471	0.816	0.471
Primary VA	1.210	1.050	1.050
Secondary V/leg	0.855 To Neutral	0.428 To Neutral	0.740
Secondary I/leg	0.577	0.816	0.471
Secondary VA	1.480	1.050	1.050
Ripple Voltage %	18.000	4.200	4.200
Ripple Frequency	3f	6f	6f

*Root mean square values to the average dc output.
Sine-wave, infinite inductance, no transformer or rectifier losses.

$$I_1 = 0.577 I_o \quad [A] \tag{11.20}$$

$$n = \frac{V_1}{E_{line}} \tag{11.21}$$

$$I_{phase} = nI_1 \tag{11.22}$$

$$I_{line} = \sqrt{3} I_{phase} \tag{11.23}$$

The total apparent power P_t for the circuit shown in Figure 11.16 is 2.10 W.

$$\text{Primary} = 1.05 \tag{11.24}$$

$$\text{Secondary} = 1.05 \tag{11.25}$$

$$P_t = P_o(1.05)\left(\frac{1}{\eta} + 1\right) \quad [\text{W}] \tag{11.26}$$

$$V_1 = 0.428V_o \quad [\text{V}] \tag{11.27}$$

$$I_1 = 0.816I_o \quad [\text{W}] \tag{11.28}$$

The total apparent power P_t for the circuit shown in Figure 11.17 is the same as for that in Figure 11.15, 2.10 W.

$$V_1 = 0.740V_o \quad [\text{V}] \tag{11.29}$$

$$I_1 = 0.471I_o \quad [\text{W}] \tag{11.30}$$

11.11 DESIGNING FOR A GIVEN REGULATION: THREE-PHASE TRANSFORMERS

Although most transformers are designed for a given temperature rise, they can also be designed for a given regulation. The regulation and power-handling ability of a core are related to two constants, K_g and K_e:

$$VA = K_g K_e \alpha \tag{11.31}$$

where α is the regulation, %. The constant K_g is determined by the core geometry,

$$K_g = \frac{W_a A_c^2 K_u}{\text{MLT}} \tag{11.32}$$

The constant K_e is determined by the magnetic and electrical operating conditions, and may be expressed by the following equation:

Table 11.4. Coefficient K_g for three-phase laminations

Core	K_g	K_g/ϕ	$\dfrac{W_a}{2}$, cm^2	A_c, cm^2 14 mil	MLT, cm
EI-1/4	0.0462	0.0154	1.25	0.363	4.28
EI-3/8	0.264	0.088	2.02	0.817	6.13
EI-1/2	0.892	0.297	2.78	1.45	7.86
EI-9/16	2.06	0.690	4.29	1.84	8.42
EI-5/8	3.60	1.20	5.65	2.27	9.69
EI-7/8	15.0	5.00	8.47	4.45	13.4
EI-1.00	35.7	11.9	14.5	5.805	16.4
EI-1.20	56.7	18.9	11.6	8.37	17.2
EI-1.50	164	54.6	18.1	12.7	21.4
EI-1.80	431	143	26.1	18.8	25.7
EI-2.40	1842	614	46.5	33.6	34.2
EI-3.60	17232	5744	105	83.6	51.1

*Where $K_u = 0.4$.
Source: Thomas & Skinner Inc.

$$K_e = 2.86 f^2 B_m^2 \times 10^{-4} \tag{11.33}$$

The derivation of the relationship for K_g and K_e is given at the end of Chapter seven. Values for the constant K_g for three-phase laminations are given in Table 11.4.

11.12 60-Hz THREE-PHASE TRANSFORMER DESIGN: AN EXAMPLE

Specifications for a three-phase transformer design (Figure 11.16) require the following:

1. E_o, 28 V
2. I_o, 3.57
3. E_{IN}, 208-3 wire
4. f, 60-Hz three-phase
5. Temperature rise, 25°C
6. Efficiency, 90% or better

Step 1. Calculate the apparent power P_t from Equation (11.26), allowing for 1.0 V diode drop (V_d):

$$P_t = P_o(1.05)\left(\frac{1}{\eta} + 1\right) \quad [\text{W}]$$

$$P_t = I_o(E_o + 2V_d)(1.05)\left(\frac{1}{\eta} + 1\right)$$

$$P_t = (3.57)(28 + 2)(1.05)\left(\frac{1}{0.9} + 1\right)$$

$$P_t = 237 \quad [\text{W}]$$

Step 2. Calculate the area product A_p from Equation (11.11):

$$A_p = \left(\frac{P_t \times 10^4}{4.44 B_m f K_u K_j}\right)^{1.14} \quad [\text{cm}^4]$$

$B_m = 1.1^*$ [T]
$K_u = 0.4$ (see Chapter 6)
$K_j = 304$ (Table 11.1)

$$A_p = \left(\frac{237 \times 10^4}{(4.44)(1.1)(60)(0.4)(304)}\right)^{1.14}$$

$$A_p = 119 \quad [\text{cm}^4]$$

After the A_p has been determined, the geometry of the transformer can be evaluated for volume in Figure 11.10, weight in Figure 11.11, and surface area in Figure 11.13, and appropriate changes made if required.

Step 3. Select a lamination from Table 11.2 with a value of A_p closest to the one calculated:

$$\text{EI-7/8}, A_p = 113 \quad [\text{cm}^4]$$

---

*This flux density was chosen to reduce stray magnetic field.

Step 4.   Calculate the total transformer losses P_Σ:

$$P_\Sigma = \frac{P_o}{\eta} - P_o \quad [\text{W}]$$

$$P_o = 107 \quad [\text{W}]$$

$$P_\Sigma = \frac{107}{0.90} - 107$$

$$P_\Sigma = 11.9 \quad [\text{W}]$$

When designing small 60-Hz transformers it is quite difficult to equalize the power losses in the core and coil; because of the large number of turns, the losses in the coil are usually much higher.

Maximum efficiency is realized when the copper (winding) losses are equal to the iron (core) losses (see Chapter 7),

$$P_{cu} = P_{fe}$$

and therefore

$$P_{cu} = \frac{P_\Sigma}{2}$$

and thus

$$P_{cu} = \frac{11.9}{2}$$

$$P_{cu} = 5.95 \quad [\text{W}]$$

Step 5.   Select the core weight from Table 11.2, column 14, and then calculate the core loss in milliwatts per gram:

$$\text{E1-7/8, } W_t = 1835 \quad [\text{g}]$$

$$\frac{P_{fe}}{W_t} \times 10^3 = mW/g$$

$$\frac{5.92}{1835} \times 10^3 = mW/g$$

$$3.23 \ mW/g$$

Step 6. Verify from the core loss curves (Figure 7.11) that 12-mil silicon at a flux density of 1.1 T will meet the requirement of 3.23 mW/g. It does, with approximately 1.5 mW/g.

Step 7. Calculate the number of primary turns using Faraday's law:

$$N_p = \frac{E_{line} \times 10^4}{4.44 B_m A_c f}$$

The iron cross section A_c is found in Table 11.2, column 16:

$$A_c = 4.45 \quad [cm^2]$$

$$N_p = \frac{208 \times 10^4}{(4.44)(1.1)(4.45)(60)} \quad [turns]$$

$$N_p = 1600 \quad [turns \ (primary)]$$

Step 8. Calculate the secondary phase voltage using Equation (11.27):

$$V_1 = 0.428 V_o$$

$$V_1 = (0.428)(30)$$

$$V_1 = 12.8 \quad [V]$$

Step 9. Calculate the secondary phase current using Equation (11.28):

$$I_1 = 0.816I_o$$

$$I_1 = (0.816)(3.57)$$

$$I_1 = 2.91 \quad [\text{A}]$$

Step 10. Calculate the turns ratio using Equation (11.21):

$$n = \frac{V_1}{E_{\text{line}}}$$

$$n = \frac{12.8}{208}$$

$$n = 0.0615$$

Step 11. Calculate the primary phase current using Equation (11.22):

$$I_{\text{phase}} = (\text{turns ratio})(I_1)$$

$$I_{\text{phase}} = (0.0615)(2.91)$$

$$I_{\text{phase}} = 0.179 \quad [\text{A}]$$

Step 12. Calculate the current density J from Equation (11.10):

$$J = K_j A_p^{-0.12}$$

Using the value for K_j from Table 11.1,

$$J = (304)(113)^{-0.12}$$

$$J = 172 \quad [\text{A/cm}^2]$$

Step 13. Calculate the primary wire size $A_{w(B)}$:

$$A_{w(B)} = \frac{I_{\text{phase}}}{J} \quad [\text{cm}^2]$$

$$A_{w(B)} = \frac{0.179}{172}$$

$$A_{w(B)} = 0.00104 \quad [\text{cm}^2]$$

Step 14. Select the wire area $A_{w(B)}$ in Table 6.1 for equivalent (AWG) wire size, column A:

$$\text{AWG No. 27} = 0.001021 \quad [\text{cm}^2]$$

The rule is that when the calculated wire size does not fall close to those listed in the table, the next smaller size should be selected.

Step 15. Calculate the resistance of the primary winding, using Table 6.1, column C, and Table 11.2, column 4, for the MLT:

$$R_p = \text{MLT} \times N \times (\text{column C}) \times \zeta \times 10^{-6} \quad [\Omega]$$

$$R_p = (13.4)(1600)(1687)(1.098) \times 10^{-6}$$

$$R_p = 39.7 \quad [\Omega]$$

Step 16. Calculate the total primary copper loss P_{cu}:

$$P_{\text{cu}} = 3(I_{\text{phase}}^2 R_p) \quad [\text{W}]$$

$$P_{\text{cu}} = 3[(0.179)^2(39.7)]$$

$$P_{\text{cu}} = 3.82 \quad [\text{W}]$$

Step 17.   Calculate the secondary turns:

$$N_s = nN_p \quad \text{[turns]}$$

$$N_s = (0.0615)(1600)$$

$$N_s = 100 \quad \text{[turns (secondary)]}$$

Step 18.   Calculate the wire size $A_{w(B)}$ for the secondary winding:

$$A_{w(B)} = \frac{I_1}{J} \quad \text{[cm}^2\text{]}$$

$$A_{w(B)} = \frac{2.91}{172}$$

$$A_{w(B)} = 0.0169 \quad \text{[cm}^2\text{]}$$

Step 19.   Select the bare wire area $A_{w(B)}$ in Table 6.1 for equivalent AWG wire size, column A:

$$\text{AWG No. 15} = 0.0165 \quad \text{[cm}^2\text{]}$$

The rule is that when the calculated wire size does not fall close to those listed in the table, the next smaller size should be selected.

Step 20.   Calculate the resistance of the secondary winding, using Table 6.1, column C, and Table 11.2, column 4, for the MLT:

$$R_s = \text{MLT} \times N \times (\text{column C}) \times \zeta \times 10^{-6} \quad \text{[}\Omega\text{]}$$

$$R_s = (13.4)(100)(104)(1.098) \times 10^{-6}$$

$$R_s = 0.153 \quad \text{[}\Omega\text{]}$$

Step 21. Calculate the total secondary copper loss P_{cu}:

$$P_{cu} = 3(I_1^2 R_s) \quad [\text{W}]$$

$$P_{cu} = 3[(2.91)^2(0.153)]$$

$$P_{cu} = 3.88 \quad [\text{W}]$$

Step 22. Summarize the losses and compare with the total losses P_Σ:

$$\text{Primary } P_{cu} = 3.82 \quad [\text{W}]$$

$$\text{Secondary } P_{cu} = 3.88 \quad [\text{W}]$$

$$\text{Core } P_{fe} = 2.75 \quad [\text{W}]$$

$$\text{Total } P_\Sigma = 10.5 \quad [\text{W}]$$

The total power loss in the transformer is 10.5 W, which will meet the required 90% efficiency.

From Chapter 7, the surface area A_t required to dissipate waste heat (expressed as watts loss per unit area) is

$$A_t = \frac{P_\Sigma}{\psi}$$

when

$$\psi = 0.03 \text{ W/cm}^2 \text{ at } 25°\text{C rise}$$

Referring to Table 11.2, column 1, for the E1-7/8 size lamination, we find surface area $A_t = 550$ cm^2, so

$$\psi = \frac{P_\Sigma}{A_t}$$

$$\psi = \frac{10.5}{556}$$

$$\psi = 0.0189 \quad [\text{W/cm}^2]$$

which will produce a temperature rise within the 25°C specification.

Step 23. Calculate the transformer regulation using Equations (11.31)–(11.33):

$$\alpha = \frac{VA}{K_g K_e} \quad [\%]$$

The transformer *VA* is

$$VA = (30)(3.57)$$

$$VA = 107$$

The value for K_g can be found in Table 11.4:

$$E1 = 7/8, K_g = 15.0$$

$$K_e = 2.86 f^2 B_m^2 \times 10^{-4}$$

$$K_e = (2.86)(60)^2(1.1)^2 \times 10^{-4}$$

$$K_e = 1.25$$

$$\alpha = \frac{107}{(15.0)(1.25)}$$

$$\alpha = 5.70 \quad [\%]$$

The three-phase transformer was wound to verify these example calculations. The test results are given in Table 11.5.

Table 11.5. Magnetic electrical data

Transformer	Resistance	
	*Calculated	Measured
Primary	39.7 Ω	30 Ω
Secondary	0.153 Ω	0.19 Ω
Regulation	5.70%	6.25%

*All winding were calculated with the same MLT

REFERENCES

1. Gibbs, J. B., *Transformer Principles and Practice,* Second Edition, McGraw-Hill Book Co., Inc., New York, 1949.
2. Lee, R., *Electronic Transformer and Circuits,* Second Edition, John Wiley & Sons, New York, N.Y., 1958.
3. E. E. Staff, M.I.T., *Magnetic Circuits and Transformers,* John Wiley & Sons, Inc., 1943.
4. *Electrical Lamination* Catalog No. ML-531, Thomas & Skinner, Inc., Indianapolis, Ind., undated.

Chapter 12

Magnetic Component Test Circuits

12.1 PROPER CONNECTION OF THE WINDING OF MAGNETIC COMPONENTS

Proper connection of the windings of magnetic components, commonly referred to as *phasing,* is an important test for proper operation. Figures 12.1–12.5 show the proper phasing of several magnetic components. The leads to all windings having the same phase or polarity are marked by a dot (•) in these diagrams.

12.2 PHASE TEST

The phasing test circuit is shown in Figure 12.6. The applied voltage E_{ac} should not be high enough to saturate the core or to put undue voltage stress on any of the windings. The test lead J1 is the

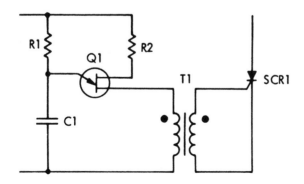

Figure 12.1. Pulse transformer.

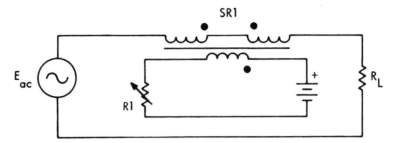

Figure 12.2. Saturable reactor.

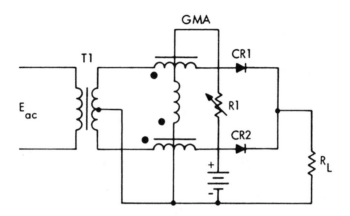

Figure 12.3. Gated magnetic amplifier.

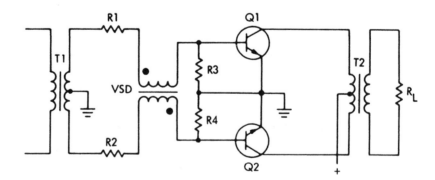

Figure 12.4. Volt-second device.

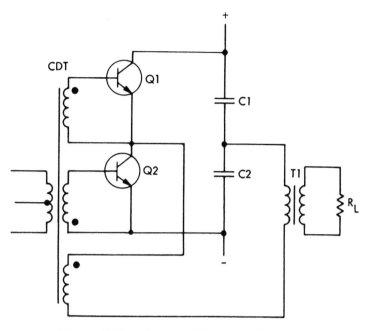

Figure 12.5. Current drive transformers.

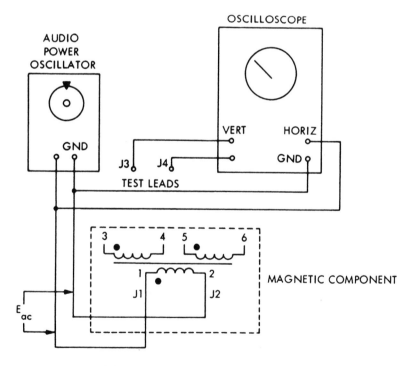

Figure 12.6. Phase test fixture.

reference voltage lead. Connect J3 of the oscilloscope to J1 of the oscillator and observe the angle of the trace; this will be the angle of the trace for all windings that are phased correctly. When J3 and J4 are connected to a winding, the correct polarity/phase will be shown in the trace on the scope.

Test lead J3 must always be connected to the start of a winding to produce the same trace angle. If the scope shows a 180° shift of the trace, the phasing of the winding under test is reversed.

12.3 MEASUREMENT OF INDUCTANCE

A convenient method of measuring inductance based on the current and voltage drop across the inductor is shown in Figure 12.7. The advantage to this method is that the inductance can be measured at a flux density close to or similar to that in an actual operating circuit. Measurements using an impedance bridge are usually done with very small currents.

The amount of applied voltage V_2 to the inductor should be limited so that the flux is operating within the linear portion of the *B-H* loop as shown in Figure 12.8.

Step 1.   Calculate the applied voltage required to operate at $B_s/2$:

$$V_2 = 4.44 \frac{B_s}{2} A_c f N \times 10^{-4}$$

(12.1)

Step 2.   Calculate the reactance:

$$X_L = 2\pi f L$$

(12.2)

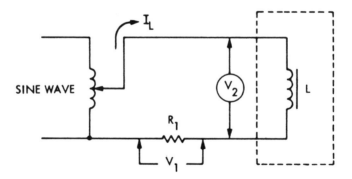

Figure 12.7.   Inductor current measurement.

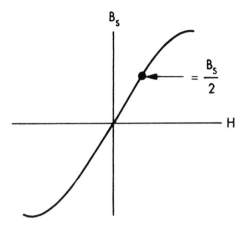

Figure 12.8. Operating flux level.

Step 3. Calculate the inductor current:

$$I_L = \frac{V_2}{X_L} \qquad (12.3)$$

Step 4. Calculate the current-sensing resistor:

$$R_1 = \frac{V_2}{I_L} \times \frac{1}{100} \qquad (12.4)$$

The voltage V_2 compared to V_1 should be very large so V_2 looks like a voltage source, $V_2 \gg V_1$. When selecting a value for R_1, it is wise to round off and select a resistor for ease of reading such as 0.1 Ω, 1.0 Ω, or 10 Ω.

Step 5. Calculate the inductor current in the reactor:

$$I_L = \frac{V_1}{R_1} \qquad (12.5)$$

Step 6. Calculate the reactance of the reactor:

$$X_L = \frac{V_2}{I_L} \qquad (12.6)$$

Step 7. Calculate the measured inductance:

$$L = \frac{X_L}{2\pi f} \qquad (12.7)$$

12.4 TESTING THE DYNAMIC *B-H* LOOP

The dynamic hysterestis or *B-H* loop contains very important information about the magnetic component. The area within the *B-H* loop relates to losses, and the amplitude relates to flux density as shown in Figure 12.9.

The circuit that is most commonly used to display the *B-H* loop on an oscilloscope is shown in Figure 12.10. The excitation current I_m causes a magnetic force in the magnetic component:

$$H = \frac{I_m N}{l_m} \quad \text{[amp-turns/cm]} \qquad (12.8)$$

The voltage drop across R_2 is proportional to the excitation current as long as the current through R_1 and C_1 is

$$I_{RC} = \frac{I_m}{100} \quad \text{[A]} \qquad (12.9)$$

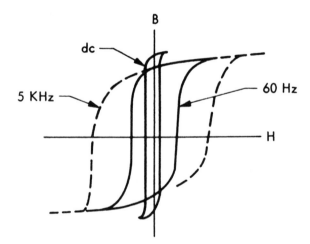

Figure 12.9. Typical *B-H* loop.

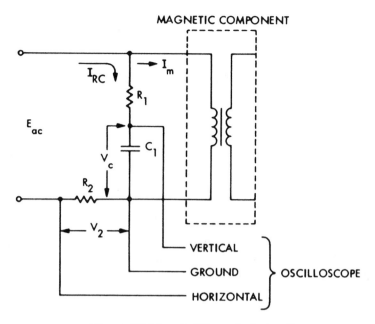

Figure 12.10. *B-H* loop test circuit.

The voltage drop V_2 should be very small compared to E_{ac}:

$$V_2 = \frac{E_{ac}}{100} \quad [\text{V}] \tag{12.9}$$

Then

$$R_2 = \frac{E_{ac}}{100I_m} \quad [\Omega] \tag{12.10}$$

The series network of R_1 and C_1 performs the integration of the applied voltage. The resistance should be very large compared to the impedance of the capacitor at the operating frequency.

$$R_1 = \frac{100}{\omega C} \quad [\Omega] \tag{12.11}$$

Then

$$R_1 = \frac{100E_{ac}}{I_m} \quad [\Omega] \tag{12.12}$$

Measurement of the *B-H* loop then is given by

$$B_m = \frac{V_c \times 10^4}{4.44 N A_c f} \quad [T] \qquad (12.13)$$

and

$$H = \frac{V_2 N}{l_m R_2} \quad [amp\text{-}turns/cm] \qquad (12.14)$$

The voltage V_c across the capacitor C_1 is directly proportional to the flux B_m. The voltage V_2 across the resistor R_2 is directly proportional to H in ampere = turns per centimeter. The oscilloscope will have to be calibrated with a known magnetic component.

12.5 TRANSFORMER REGULATION TEST

A transformer regulation test can be carried out using the test circuit shown in Figure 12.11 and without high voltages. Regulation has to do only with the resistance of that winding and the current through it.

Step 1. Place transformer in the test fixture and close S1.

Step 2. Adjust V_1 until I_o is equal to nominal load current, and record V_1.

Step 3. Open S1 and readjust V_1 to the value recorded in step 2, and then record the reading on V_2.

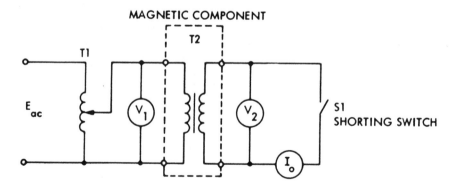

Figure 12.11. Transformer regulation test circuit.

Step 4. Adjust V_1 to nominal input voltage and then record the reading on V_2. The regulation (α) of the transformer is then:

$$\alpha = \frac{(V_2 \text{ nominal}) - (V_2 \text{ after short})}{V_2 \text{ nominal}} \times 100 \quad [\%] \tag{12.15}$$

12.6 TRANSFORMER DESIGN USING THE CORE GEOMETRY K_g APPROACH

Design work sheets for the following specifications:

1. Input voltage, $V_{in} = ($ $)$ [V]
2. Output voltage, $V_o = ($ $)$ [V]
3. Output current, $I_o = ($ $)$ [A]
4. Frequency, $f = ($ $)$ [Hz]
5. Efficiency, $\eta = ($ $)$
6. Regulation, $\alpha = ($ $)$ [%]
7. Flux density, $B_m = ($ $)$ [T]
8. Core material: _____
9. Core configuration: _____

Step 1. (See design note 1.)* Calculate the transformer output power P_o, allowing for an assumed 1.0 V diode drop V_d:

$$P_o = VA \quad [W]$$

$$V = V_o + V_d \quad [V]$$

$$V = (\quad) + (\quad)$$

$$P_o = (\quad)(\quad)$$

$$P_o = \qquad [W]$$

Step 2. (See design note 2.) Calculate the apparent power P_t:

$$P_t = P_o\left(\frac{\sqrt{2}}{\eta} + \sqrt{2}\right) \quad [W]$$

---

*Design notes appear on page 412.

$$P_t = (\quad)\left(\frac{1.41}{(\quad)} + 1.41\right)$$

$$P_t = \qquad [W]$$

Step 3. (See design note 3.) Calculate the electrical conditions:

$$K_e = 0.145 K_f^2 f^2 B_m^2 \times 10^{-4}$$

$$K_f =$$

$$K_e = 0.145(\quad)^2(\quad)^2(\quad)^2 \times 10^{-4}$$

$$K_e =$$

Step 4. Calculate the core geometry K_g:

$$K_g = \frac{P_t}{2 K_e \alpha} \quad [cm^5]$$

$$K_g = \frac{(\quad)}{(\quad)(\quad)}$$

$$K_e = \qquad [cm^5]$$

Step 5. (See design note 4.) Select a comparable core geometry K_g from the specified core section, and record the appropriate data:

$$\text{Core} \underline{\qquad}$$

$$K_g = \qquad [cm^5]$$

$$A_p = \qquad [cm^4]$$

$$MLT = \quad\quad [cm]$$

$$A_c = \quad\quad [cm^2]$$

$$W_a = \quad\quad [cm^2]$$

$$A_t = \quad\quad [cm^2]$$

$$MPL = \quad\quad [cm]$$

$$W_{tfe} = \quad\quad [kg]$$

Step 6. Calculate the number of primary turns each side of the center tap:

$$N_p = \frac{V_p \times 10^4}{K_f B_m f A_c} \quad [turns]$$

$$N_p = \frac{(\quad) \times 10^4}{(\quad)(\quad)(\quad)(\quad)}$$

$$N_p = \quad\quad [turns]$$

Step 7. Calculate the primary current I_p:

$$I_p = \frac{P_o}{V_p \eta} \quad [A]$$

$$I_p = \frac{(\quad)}{(\quad)(\quad)}$$

$$I_p = \quad\quad [A]$$

Step 8. Calculate the current density J:

$$J = \frac{P_t \times 10^4}{K_f K_u f B_m A_p} \quad [\text{A/cm}^2]$$

Window utilization factor $K_u = 0.4$

$$J = \frac{(\quad) \times 10^4}{(\quad)(\quad)(\quad)(\quad)(\quad)}$$

$$J = \qquad [\text{A/cm}^2]$$

Step 9. (See design note 5.) Calculate the bare wire size $A_{w(B)}$ for the primary:

$$A_{w(B)} = \frac{I_p}{J} \quad [\text{cm}^2]$$

$$A_{w(B)} = \frac{(\quad)}{(\quad)}$$

$$A_{w(B)} = \qquad [\text{cm}^2]$$

Step 10. Select a wire size from the wire table (Table 6.1), column 2. If the wire area is not within 10%, take the next smallest size. Also record micro-ohms per centimeter from column 4.

AWG No. ____

Bare, $A_{w(B)} =$ ____ $[\text{cm}^2]$

$\mu\Omega/\text{cm} =$

Step 11. Calculate the primary winding resistance. Use MLT from step 5 and micro-ohms per centimeter from step 10:

$$R_p = (\text{MLT})(N)\left(\frac{\mu\Omega}{\text{cm}}\right) \times 10^{-6} \quad [\Omega]$$

$$R_p = (\quad)(\quad)(\quad) \times 10^{-6}$$

$$R_p = \qquad [\Omega]$$

Step 12. Calculate the primary copper loss P_p:

$$P_p = (I_p)^2 R_p \quad [W]$$

$$P_p = (\quad)^2(\quad)$$

$$P_p = \qquad [W]$$

Step 13. Calculate the secondary turns each side of the center tap:

$$N_s = \left(\frac{N_p V_s}{V_p}\right)\left(1 + \frac{\alpha}{100}\right) \quad [\text{turns}]$$

$$N_s = \frac{(\quad)(\quad)}{(\quad)}\left(1 + \frac{(\quad)}{100}\right)$$

$$N_s = \qquad [\text{turns}]$$

Step 14. (See design note 5.) Calculate the bar wire size $A_{w(B)}$ for the secondary.

$$A_{w(B)} = \frac{I_o}{J} \quad [\text{cm}^2]$$

$$A_{w(B)} = \frac{(\quad)}{(\quad)}$$

$$A_{w(B)} = \qquad [\text{cm}^2]$$

Step 15. Select a wire size from the wire table (Table 6.1), column 2. If the wire area is not within 10%, take the next smallest size. Also record micro-ohms per centimeter from column 4.

AWG No. ____

Bare, $A_{w(B)} =$ [cm^2]

$\mu\Omega/\text{cm} =$

Step 16.   Calculate the secondary winding resistance. Use MLT from step 5 and micro-ohms per centimeter from step 15:

$$R_s = (\text{MLT})(N)\left(\frac{\mu\Omega}{\text{cm}}\right) \times 10^{-6} \quad [\Omega]$$

$$R_s = (\quad)(\quad)(\quad) \times 10^{-6}$$

$$R_s = \qquad [\Omega]$$

Step 17.   Calculate the secondary copper loss P_s:

$$P_s = I_o^2 R_s \quad [\text{W}]$$

$$P_s = (\quad)^2(\quad)$$

$$P_s = \qquad [\text{W}]$$

Step 18.   Calculate the transformer regulation, α:

$$\alpha = \frac{P_{cu} \times 100}{P_o} \quad [\%]$$

$$P_{cu} = P_p + P_s \quad [\text{W}]$$

$$P_{cu} = (\quad) + (\quad)$$

$$P_{cu} = \qquad [\text{W}]$$

$$\alpha = \frac{(\quad) \times 100}{(\quad)}$$

$$\alpha = \qquad [\%]$$

Step 19. Calculate what the combined losses P_Σ have to be to meet the efficiency specification:

$$P_\Sigma = \frac{P_o}{\eta} - P_o \quad [\text{W}]$$

$$P_\Sigma = \frac{(\quad)}{(\quad)} - (\quad)$$

$$P_\Sigma = \qquad [\text{W}]$$

Step 20. Calculate the iron losses P_{fe}:

$$P_{\text{fe}} = P_\Sigma - P_{\text{cu}} \quad [\text{W}]$$

$$P_{\text{fe}} = (\quad) - (\quad)$$

$$P_{\text{fe}} = \qquad [\text{W}]$$

Step 21. Calculate the core loss P_{fe} in milliwatts per gram:

$$\text{mW/g} = \frac{P_{\text{fe}}}{W_{\text{tfe}} \times 10^{-3}}$$

$$\text{mW/g} = \frac{(\quad)}{(\quad) \times 10^{-3}}$$

$$\text{mW/g} =$$

Step 22. (See design note 9.) Using the appropriate core loss curves, find the magnetic material that comes the closest to meeting the requirement. Then calculate the watts per kilogram. Core weight is found in step 4.

$$W/kg = K f^{(m)} B_m^{(n)}$$

$$W/kg = (\quad)(\quad)^{(\ )}(\quad)^{(\ )} \times 10^{-3}$$

$$W/kg =$$

$$P_{fe} = (W/kg)(W_{tfe}) \quad [W]$$

$$P_{fe} = (\quad)(\quad)$$

$$P_{fe} = \qquad [W]$$

12.7 TOROIDAL INDUCTOR DESIGN USING THE CORE GEOMETRY K_g APPROACH

Design work sheets with the following specifications:

1. Inductance, $L = (\quad)$ [H]
2. dc current, $I_o = (\quad)$ [A]
3. ac current, $\Delta I = (\quad)$ [A]
4. Output power, $P_o = (\quad)$ [W]
5. Regulation, $\alpha = (\quad)$ [%]
6. Ripple frequency, $f = (\quad)$ [Hz]
7. Flux density, $B_m = (\quad)$ [T]
8. Core material: _____
9. Core configuration: _____

Step 1. Calculate the energy-handling capability:

$$\text{Energy} = \frac{LI^2}{2} \quad [W\text{-}s]$$

$$I = I_o + \frac{\Delta I}{2} \quad [A]$$

$$I = (\quad) + \frac{(\quad)}{2}$$

$$I = \qquad [A]$$

$$\text{Energy} = \frac{(\quad)(\quad)^2}{2}$$

$$\text{Energy} = \qquad [\text{W-s}]$$

Step 2. (See design note 6.)* Calculate the electrical conditions parameter K_e:

$$K_e = 0.145 P_o B_m^2 \times 10^{-4}$$

$$K_e = 0.145(\quad)(\quad)^2 \times 10^{-4}$$

$$K_e =$$

Step 3. Calculate the core geometry coefficient K_g:

$$K_g = \frac{(\text{Energy})^2}{K_e \alpha} \quad [\text{cm}^5]$$

$$K_g = \frac{(\quad)^2}{(\quad)(\quad)}$$

$$K_g = \qquad [\text{cm}^5]$$

Step 4. (See design note 4.) Select a comparable core geometry K_g from the specified core section, and record the appropriate data.

$$\text{Core} \underline{\qquad}$$

$$K_g = \qquad [\text{cm}^5]$$

---

*Design notes appear on page 412.

$$A_p = \qquad [\text{cm}^4]$$

$$\text{MLT} = \qquad [\text{cm}]$$

$$A_c = \qquad [\text{cm}^2]$$

$$W_a = \qquad [\text{cm}^2]$$

$$A_t = \qquad [\text{cm}^2]$$

$$\text{MPL} = \qquad [\text{cm}]$$

$$W_{tfe} = \qquad [\text{g}]$$

Step 5. Calculate the current density J. Use area product A_p found in step 4:

$$J = \frac{2(\text{Energy}) \times 10^4}{B_m A_p K_u} \quad [\text{A/cm}^2]$$

Window utilization factor $K_u = 0.4$

$$J = \frac{2(\quad) \times 10^4}{(\quad)(\quad)(\quad)}$$

$$J = \qquad [\text{A/cm}^2]$$

Step 6. Calculate the bare wire size $A_{w(B)}$:

$$A_{w(B)} = \frac{I_o + \Delta I/2}{J} \quad [\text{cm}^2]$$

$$A_{w(B)} = \frac{(\quad) + (\quad)/2}{(\quad)}$$

$$A_{w(B)} = \qquad [\text{cm}^2]$$

Step 7. Select a wire size from the wire table (Table 6.1), column 2. If the area is not within 10%, take the next smallest size. Also record micro-ohms per centimeter from column 4.

AWG No. ____

Bare, $A_{w(B)} = \qquad [\text{cm}^2]$

Insulated, $A_w = \qquad [\text{cm}^2]$

$\mu\Omega/\text{cm} = $

Step 8. Calculate the effective window area $W_{a(\text{eff})}$. Use window area W_a found in step 4:

$$W_{a(\text{eff})} = W_a S_3 \quad [\text{cm}^2]$$

A typical value for S_3 is 0.75, as shown in Chapter 6.

$$W_{a(\text{eff})} = (\quad)(0.75)$$

$$W_{a(\text{eff})} = \qquad [\text{cm}^2]$$

Step 9. Calculate N, the number of turns. Use wire area A_w found in step 7:

$$N = \frac{W_{a(\text{eff})} S_2}{A_w} \quad [\text{turns}]$$

A typical value for S_2 is 0.6, as shown in Chapter 6.

$$N = \frac{(\quad)(0.6)}{(\quad)}$$

$$N = \qquad [\text{turns}]$$

Step 10.   Calculate the permeability of the core required:

$$\mu_r = \frac{L \times \text{MPL} \times 10^8}{0.4\pi N^2 A_c}$$

$$\mu_r = \frac{(\quad)(\quad) \times 10^8}{(1.26)(\quad)^2(\quad)}$$

$$\mu_r =$$

Now choose a core from the group with a closer permeability.

No. _____

$$\mu =$$

$$\text{MH}/1000 =$$

Step 11.   Calculate *N*, the number of turns required:

$$N = 1000 \sqrt{\frac{L}{L_{1000}}} \quad [\text{turns}]$$

$$N = 1000 \sqrt{\frac{(\quad)}{(\quad)}}$$

$$N = \qquad [\text{turns}]$$

Step 12.   Calculate the winding resistance. Use MLT from step 4 and micro-ohms per centimeter from step 7.

$$R = (\text{MLT})(N)\left(\frac{\mu\Omega}{\text{cm}}\right) \times 10^{-6} \quad [\Omega]$$

$$R = (\quad)(\quad)(\quad) \times 10^{-6}$$

$$R = \qquad [\Omega]$$

Step 13. Calculate the copper loss P_{cu}:

$$P_{cu} = I^2 R \quad [W]$$

$$P_{cu} = (\quad)^2(\quad)$$

$$P_{cu} = \qquad [W]$$

Step 14. Calculate the regulation, α:

$$\alpha = \frac{P_{cu}}{P_o} \times 100 \quad [\%]$$

$$\alpha = \frac{(\quad) \times 100}{(\quad)}$$

$$\alpha = \qquad [\%]$$

Step 15. Calculate the ac flux density:

$$B_{ac} = \frac{0.4\pi N(\Delta I/2)\mu \times 10^{-4}}{\text{MPL}} \quad [T]$$

$$B_{ac} = \frac{(1.26)(\quad)(\quad)(\quad) \times 10^{-4}}{(\quad)}$$

$$B_{ac} = \qquad [T]$$

Step 16. (See design note 9.) Calculate the watts per kilogram for the appropriate core material, and then determine the core loss. Core weight is found in step 4.

$$\text{W/kg} = K f^{(m)} B_m^{(n)}$$

$$mW/g = (\quad)(\quad)^{(\ )}(\quad)^{(\ )}$$

$$mW/g =$$

$$P_{fe} = (mW/g)(W_{tfe}) \times 10^{-3} \quad [W]$$

$$P_{fe} = (\quad)(\quad) \times 10^{-3}$$

$$P_{fe} = \qquad [W]$$

Step 17. Calculate the total losses P_{Σ}:

$$P_{\Sigma} = P_{cu} + P_{fe} \quad [W]$$

$$P_{\Sigma} = (\quad) + (\quad)$$

$$P_{\Sigma} = \qquad [W]$$

Step 18. Calculate the watts per unit area. The surface area A_t is found in step 4:

$$\psi = \frac{P_{cu}}{A_t} \quad [W/cm^2]$$

$$\psi = \frac{(\quad)}{(\quad)}$$

$$\psi = \qquad [W/cm^2]$$

Step 19. (See design note 7.) Calculate the dc magnetizing force:

$$H = \frac{0.4\pi NI}{MPL} \quad [Oe]$$

$$H = \frac{(1.26)(\quad)(\quad)}{(\quad)}$$

$$H = \qquad \text{[Oe]}$$

$$NI = (H)(0.8) \quad \text{[amp-turns]}$$

$$NI = (\quad)(0.8) \quad \text{[amp-turns]}$$

12.8 GAPPED INDUCTOR DESIGN USING THE CORE GEOMETRY K_g APPROACH

Design work sheets for a gapped inductor with the following specifications.

1. Inductance, $L = (\quad)$ [H]
2. dc current, $I_o = (\quad)$ [A]
3. ac current, $\Delta I = (\quad)$ [A]
4. Output power, $P_o = (\quad)$ [W]
5. Regulation, $\alpha = (\quad)$ [%]
6. Ripple frequency, $f = (\quad)$ [Hz]
7. Flux density, $B_m = (\quad)$ [T]
8. Core material: _____
9. Core configuration: _____

Step 1. Calculate the energy-handling capability:

$$\text{Energy} = \frac{LI^2}{2} \quad \text{[W-s]}$$

$$I = I_o + \frac{\Delta I}{2} \quad \text{[A]}$$

$$I = (\quad) + \frac{(\quad)}{2}$$

$$I = \qquad \text{[A]}$$

$$\text{Energy} = \frac{(\quad)(\quad)^2}{2}$$

$$\text{Energy} = \qquad \text{[W-s]}$$

Step 2. Calculate the electrical conditions parameter K_e:

$$K_e = 0.145 P_o B_m^2 \times 10^{-4}$$

$$K_e = (0.145)(\quad)(\quad)^2 \times 10^{-4}$$

$$K_e =$$

Step 3. Calculate the core geometry coefficient K_g:

$$K_g = \frac{(\text{Energy})^2}{K_e \alpha} \quad \text{[cm}^5\text{]}$$

$$K_g = \frac{(\quad)^2}{(\quad)(\quad)}$$

$$K_g = \qquad \text{[cm}^5\text{]}$$

Step 4. (See design note 4.)* Select a comparable core geometry K_g from the specified core section, and record the appropriate data.

$$\text{Core} \underline{\qquad}$$

$$K_g = \qquad \text{[cm}^5\text{]}$$

$$A_p = \qquad \text{[cm}^4\text{]}$$

$$\text{MLT} = \qquad \text{[cm]}$$

$$A_c = \qquad \text{[cm}^2\text{]}$$

---

*Design notes appear on page 412.

$$W_a = \qquad [cm^2]$$

$$A_t = \qquad [cm^2]$$

$$G = \qquad [cm]$$

$$MPL = \qquad [cm]$$

$$W_{tfe} = \qquad [g]$$

Step 5. Calculate the current density J. Use area product A_p found in step 4:

$$J = \frac{2(\text{Energy}) \times 10^4}{B_m A_p K_u} \quad [A/cm^2]$$

Window utilization factor $K_u = 0.4$

$$J = \frac{2(\quad) \times 10^4}{(\quad)(\quad)(\quad)}$$

$$J = \qquad [A/cm^2]$$

Step 6. Calculate the bare wire size $A_{w(B)}$:

$$A_{w(B)} = \frac{I_o + \Delta I/2}{J} \quad [cm^2]$$

$$A_{w(B)} = \frac{(\quad)(\quad)}{(\quad)}$$

$$A_{w(B)} = \qquad [cm^2]$$

Step 7. Select a wire size from the wire table (Table 6.1), column 2. If the area is not within 10%, take the next smallest size. Also record micro-ohms per centimeter from column 4 and wire area with insulation A_w from column 5.

AWG No. ____

Bare, $A_{w(B)}$ =   [cm^2]

Insulated, A_w =   [cm^2]

$\mu\Omega$/cm =

Step 8. Calculate the effective window area $W_{a(\text{eff})}$. Use the window area W_a found in step 4:

$$W_{a(\text{eff})} = W_a S_3 \quad [\text{cm}^2]$$

A typical value for S_3 is 0.75, as shown in Chapter 6.

$$W_{a(\text{eff})} = (\quad)(0.75)$$

$$W_{a(\text{eff})} = \qquad [\text{cm}^2]$$

Step 9. Calculate N, the number of turns. Use wire area A_w found in step 7:

$$N = \frac{W_{a(\text{eff})} S_2}{A_w} \quad [\text{turns}]$$

A typical value for S_2 is 0.6, as shown in Chapter 6.

$$N = \frac{(\quad)(\quad)}{(\quad)}$$

$$N = \qquad [\text{turns}]$$

Step 10. Calculate the required gap, and use iron area A_c found in step 4:

$$l_g = \frac{0.4N^2 A_c \times 10^{-8}}{L} \quad [\text{cm}]$$

$$l_g = \frac{1.26(\quad)^2(\quad) \times 10^{-8}}{(\quad)}$$

$$l_g = \quad [\text{cm}]$$

Gap spacing is usually maintained by inserting kraft paper. However, this paper is available only in mil thicknesses. Since l_g has been determined in cm, it is necessary to convert as follows:

$$\text{cm} \times 393.7 = \text{mils}$$

Substituting values:

$$(\quad)(393.7) = \quad [\text{mils}]$$

Round off to nearest even mil and multiply:

$$(\quad)2.54 = \quad [\text{cm}^2]$$

Step 11. Calculate the amount of fringing flux. Use the G dimension recorded in step 4:

$$F = 1 + \frac{l_g}{\sqrt{A_c}} \ln\left(\frac{2G}{l_g}\right)$$

$$F = 1 + \frac{(\quad)}{(\quad)} \ln\left(\frac{2(\quad)}{(\quad)}\right)$$

$$F = $$

Step 12. Calculate the new number of turns by inserting the fringing flux:

$$N = \left(\frac{l_g L}{0.4\pi A_c F \times 10^{-8}}\right)^{1/2} \quad [\text{turns}]$$

$$N = \left(\frac{(\quad)(\quad)}{(1.26)(\quad)(\quad) \times 10^{-8}} \right)^{1/2}$$

$$N = \qquad \text{[turns]}$$

Step 13. Calculate the winding resistance. Use MLT from step 4 and micro-ohms per centimeter from step 7.

$$R = (\text{MLT})(N)\left(\frac{\mu\Omega}{\text{cm}} \right) \times 10^{-6} \quad [\Omega]$$

$$R = (\quad)(\quad)(\quad) \times 10^{-6}$$

$$R = \qquad [\Omega]$$

Step 14. Calculate the copper loss P_{cu}:

$$P_{cu} = I^2 R \quad [\text{W}]$$

$$P_{cu} = (\quad)^2(\quad)$$

$$P_{cu} = \qquad [\text{W}]$$

Step 15. Calculate the regulation, α:

$$\alpha = \frac{P_{cu}}{P_o} \times 100 \quad [\%]$$

$$\alpha = \frac{(\quad) \times 100}{(\quad)}$$

$$\alpha = \qquad [\%]$$

Step 16. Calculate the total ac plus dc flux density:

$$B_m = \frac{0.4\pi N(I_{dc} + \Delta I/2) \times 10^{-4}}{(\quad)} \quad [\text{T}]$$

$$B_m = \frac{(12.6)(\quad)(\quad + \quad/2) \times 10^{-4}}{(\quad)}$$

$$B_m = \quad [\text{T}]$$

Step 17. Calculate the ac flux density:

$$B_m = \frac{0.4\pi N(\Delta I/2) \times 10^{-4}}{l_g} \quad [\text{T}]$$

$$B_m = \frac{1.26(\quad)(\quad) \times 10^{-4}}{(\quad)}$$

$$B_m = \quad [\text{T}]$$

Step 18. (See design note 9.) Calculate the watts per kilogram for the appropriate core material, and then determine the core loss. Core weight is found in step 4.

$$\text{W/kg} = K f^{(m)} B_m^{(n)} \quad [\text{T}]$$

$$\text{W/kg} = (\quad)(\quad)^{(\ )}(\quad)^{(\ )}$$

$$\text{W/kg} =$$

$$P_{fe} = (\text{W/kg}) W_{tfe} \quad [\text{W}]$$

$$P_{fe} = (\quad)(\quad)$$

$$P_{fe} = \quad [\text{W}]$$

Step 19.   Calculate the losses P_Σ:

$$P_\Sigma = P_{cu} + P_{fe} \quad [W]$$

$$P_\Sigma = (\quad) + (\quad)$$

$$P_\Sigma = \qquad [W]$$

Step 20.   Calculate the efficiency of the inductor:

$$\eta = \frac{P_o \times 100}{P_o + P_\Sigma} \quad [\%]$$

$$\eta = \frac{(\quad) \times 100}{(\quad) + (\quad)}$$

$$\eta = \qquad [\%]$$

Step 21.   Calculate the watts per unit area. The surface area A_t is found in step 4.

$$\psi = \frac{P_\Sigma}{A_t} \quad [W/cm^2]$$

$$\psi = \frac{(\quad)}{(\quad)}$$

$$\psi = \qquad [W/cm^2]$$

12.9 TRANSFORMER DESIGN USING THE AREA PRODUCT A_p APPROACH

Design work sheets with the following specifications:

1. Input voltage, $V_{in} = (\quad)$ [V]
2. Output voltage, $V_o = (\quad)$ [V]
3. Output current, $I_o = (\quad)$ [A]
4. Frequency, $f = (\quad)$ [Hz]

5. Efficiency, η = (   )
6. Flux density, B_m = (   ) [T]
7. Core material: _____
8. Core configuration: _____

Step 1. (See design note 1.*) Calculate the transformer output power P_o allowing for an assumed 1.0 V diode drop V_d:

$$P_o = VA \quad [\text{W}]$$

$$V = V_o + V_d \quad [\text{V}]$$

$$V = (\quad) + (\quad)$$

$$P_o = (\quad)(\quad)$$

$$P_o = \qquad [\text{W}]$$

Step 2. (See design note 2.) Calculate the apparent power P_t:

$$P_t = P_o\left(\frac{\sqrt{2}}{\eta} + \sqrt{2}\right) \quad [\text{W}]$$

$$P_t = (\quad)\left(\frac{1.41}{(\quad)} + 1.41\right)$$

$$P_t = \qquad [\text{W}]$$

Step 3. (See design note 8.) Calculate the area product A_p:

$$A_p = \left(\frac{P_t \times 10^4}{K_f B_m f K_u K_j}\right)^{(x)} \quad [\text{cm}^4]$$

---

*Design notes appear on page 412.

Window utilization factor $K_u = 0.4$

$$A_p = \left(\frac{(\quad) \times 10^4}{(\quad)(\quad)(\quad)(\quad)(\quad)} \right)^{(\;)}$$

$$A_p = \qquad [\text{cm}^4]$$

Step 4. (See design note 4.) Select a comparable area product A_p from the specified core section, and record the appropriate data:

Core _____

$$A_p = \qquad [\text{cm}^4]$$

$$\text{MLT} = \qquad [\text{cm}]$$

$$A_c = \qquad [\text{cm}^2]$$

$$W_a = \qquad [\text{cm}^2]$$

$$A_t = \qquad [\text{cm}^2]$$

$$\text{MPL} = \qquad [\text{cm}]$$

$$W_{\text{tfe}} = \qquad [\text{kg}]$$

Step 5. Calculate the number of primary turns on each side of the center tap:

$$N_p = \frac{V_p \times 10^4}{K_f B_m f A_c} \quad [\text{turns}]$$

$$N_p = \frac{(\quad) \times 10^4}{(\quad)(\quad)(\quad)(\quad)}$$

$$N_p = \qquad \text{[turns]}$$

Step 6. Calculate the primary current I_p:

$$I_p = \frac{P_o}{V_p \eta} \quad \text{[A]}$$

$$I_p = \frac{(\qquad)}{(\qquad)(\qquad)}$$

$$I_p = \qquad \text{[A]}$$

Step 7. (See design note 8.) Calculate the current density J:

$$J = K_j A_p^{(y)} \quad \text{[A/cm}^2\text{]}$$

$$J = (\qquad)(\qquad)^{(\ )}$$

$$J = \qquad \text{[A/cm}^2\text{]}$$

Step 8. (See design note 5.) Calculate the bare wire size $A_{w(B)}$ for the primary:

$$A_{w(B)} = \frac{I_p}{J} \quad \text{[cm}^2\text{]}$$

$$A_{w(B)} = \frac{(\qquad)}{(\qquad)}$$

$$A_{w(B)} = \qquad \text{[cm}^2\text{]}$$

Step 9. Select a wire size from the wire table (Table 6.1), column 2. If the wire area is not within 10%, take the next smallest size. Also record micro-ohms per centimeter from column 4.

AWG No. ____

$$\text{Bare, } A_{w(B)} = \qquad [\text{cm}^2]$$

$$\mu\Omega/\text{cm} =$$

Step 10. Calculate the primary winding resistance. Use MLT from step 4 and micro-ohms per centimeter from step 9:

$$R_p = (\text{MLT})(N)\left(\frac{\mu\Omega}{\text{cm}}\right) \times 10^{-6} \quad [\Omega]$$

$$R_p = (\quad)(\quad)(\quad) \times 10^{-6}$$

$$R_p = \qquad [\Omega]$$

Step 11. Calculate the primary copper loss P_p:

$$P_p = I_p^2 R_p \quad [\text{W}]$$

$$P_p = (\quad)^2(\quad)$$

$$P_p = \qquad [\text{W}]$$

Step 12. Calculate the secondary turns each side of the center tap:

$$N_s = \frac{N_p V_s}{V_p} \quad [\text{turns}]$$

$$N_s = \frac{(\quad)(\quad)}{(\quad)}$$

$$N_s = \qquad [\text{turns}]$$

Step 13. (See design note 5.) Calculate the bare wire size $A_{w(B)}$ for the secondary:

$$A_{w(B)} = \frac{I_o}{J} \quad [\text{cm}^2]$$

$$A_{w(B)} = \frac{(\quad)}{(\quad)}$$

$$A_{w(B)} = \quad [\text{cm}^2]$$

Step 14. Select a wire size from the wire table (Table 6.1), column 2. If the wire area is not within 10%, take the next smallest size. Also record micro-ohms per centimeter from column 4.

AWG No. ____

Bare, $A_{w(B)} = \quad [\text{cm}^2]$

$\mu\Omega/\text{cm} =$

Step 15. Calculate the secondary winding resistance. Use MLT from step 5 and micro-ohms per centimeter from step 14:

$$R_s = (\text{MLT})(N)\left(\frac{\mu\Omega}{\text{cm}}\right) \times 10^{-6} \quad [\Omega]$$

$$R_s = (\quad)(\quad)(\quad) \times 10^{-6}$$

$$R_s = \quad [\Omega]$$

Step 16. Calculate the secondary copper loss P_s:

$$P_s = I_o^2 R_s \quad [\text{W}]$$

$$P_s = (\quad)^2(\quad)$$

$$P_s = \quad [\text{W}]$$

Step 17. Calculate the transformer total copper loss:

$$P_{cu} = P_p + P_s \quad [W]$$

$$P_{cu} = (\quad) + (\quad)$$

$$P_{cu} = \quad [W]$$

Step 18. Calculate what the combined losses P_Σ have to be to meet the efficiency specification:

$$P_\Sigma = \frac{P_o}{\eta} - P_o \quad [W]$$

$$P_\Sigma = \frac{(\quad)}{(\quad)} - (\quad)$$

$$P_\Sigma = \quad [W]$$

Step 19. Calculate the iron losses P_{fe}:

$$P_{fe} = P_\Sigma - P_{cu} \quad [W]$$

$$P_{fe} = (\quad) - (\quad)$$

$$P_{fe} = \quad [W]$$

Step 20. Calculate the core loss P_{fe} in milliwatts per gram:

$$mW/g = \frac{P_{fe}}{W_{tfe} \times 10^{-3}}$$

$$mW/g = \frac{(\quad)}{(\quad) \times 10^{-3}}$$

$$mW/g =$$

Step 21. (See design note 9.) Using the appropriate core loss curves, find the magnetic material that comes the closest to meeting the requirement. Then calculate the watts per kilogram. Core weight is found in step 4.

$$W/kg = K f^{(m)} B_m^{(n)}$$

$$W/kg = (\quad)(\quad)^{(\ )}(\quad)^{(\ )} \times 10^{-3}$$

$$W/kg =$$

$$P_{fe} = (W/kg)(W_{tfe}) \quad [W]$$

$$P_{fe} = (\quad)(\quad)$$

$$P_{fe} = \quad\quad [W]$$

Step 22. Calculate the watts per unit area. The surface area A_t is found in step 4.

$$\psi = \frac{P_{cu} + P_{fe}}{A_t} \quad [W/cm^2]$$

$$\psi = \frac{(\quad) + (\quad)}{(\quad)}$$

$$\psi = \quad\quad [W/cm^2]$$

12.10 GAPPED INDUCTOR DESIGN USING THE AREA PRODUCT A_p APPROACH

Design work sheets for a gapped inductor with the following specifications:

1. Inductance, $L = (\quad)$ [H]
2. dc current, $I_o = (\quad)$ [A]
3. ac current, $\Delta I = (\quad)$ [A]
4. Ripple frequency, $f = (\quad)$ [Hz]

5. Flux density, $B_m = ($ $)$ [T]
6. Core material: _____
7. Core configuration: _____

Step 1. Calculate the energy-handling capability:

$$\text{Energy} = \frac{LI^2}{2} \quad [\text{W-s}]$$

$$I = I_o + \frac{\Delta I}{2} \quad [\text{A}]$$

$$I = (\quad) + \frac{(\quad)}{2}$$

$$I = \quad\quad [\text{A}]$$

$$\text{Energy} = \frac{(\quad)(\quad)^2}{2}$$

$$\text{Energy} = \quad\quad [\text{W-s}]$$

Step 2. (See design note 8.)* Calculate the area product A_p:

$$A_p = \left(\frac{2(\text{Energy}) \times 10^4}{B_m K_u K_j}\right)^{(x)} \quad [\text{cm}^4]$$

Window utilization factor $K_u = 0.4$

$$A_p = \left(\frac{2(\quad) \times 10^4}{(\quad)(0.4)(\quad)}\right)^{(\quad)}$$

$$A_p = \quad\quad [\text{cm}^4]$$

_____

*Design notes appear on page 412.

Step 3. (See design note 4.) Select a comparable area product A_p from the specified core section, and record the appropriate data:

Core _____

$A_p =$ [cm⁴]

MLT =   [cm]

$A_c =$ [cm²]

$W_a =$ [cm²]

$A_t =$ [cm²]

MPL =   [cm]

$W_{tfe} =$ [g]

$G =$ [cm]

Step 4. (See design note 8.) Calculate the current density J:

$$J = K_j A_p^{(y)} \quad [\text{A/cm}^2]$$

$$J = (\quad)(\quad)^{(\ )}$$

$$J = \quad [\text{A/cm}^2]$$

Step 5. Calculate the bare wire size $A_{w(B)}$:

$$A_{w(B)} = \frac{I_o + \Delta I/2}{J} \quad [\text{cm}^2]$$

$$A_{w(B)} = \frac{(\quad) + (\quad/2)}{(\quad)}$$

$$A_{w(B)} = \qquad [cm^2]$$

Step 6. Select a wire size from the wire table (Table 6.1), column 2. If the area is not within 10%, take the next smallest size. Also record micro-ohms per centimeter from column 4 and wire area with insulation, A_w, from column 5.

AWG No. ____

Bare, $A_{w(B)} = \qquad [cm^2]$

Insulated, $A_w = \qquad [cm^2]$

$\mu\Omega/cm = $

Step 7. Calculate the effective window area $W_{a(eff)}$. Use the window area W_a found in step 3:

$$W_{a(eff)} = W_a S_3 \quad [cm^2]$$

A typical value for S_3 is 0.75, as shown in Chapter 6.

$$W_{a(eff)} = (\quad)(0.75)$$

$$W_{a(eff)} = \qquad [cm^2]$$

Step 8. Calculate N, the number of turns. Use wire area A_w found in step 6:

$$N = \frac{W_{a(eff)} S_2}{A_w} \quad [turns]$$

A typical value for S_2 is 0.6, as shown in Chapter 6.

$$N = \frac{(\quad)(\quad)}{(\quad)}$$

$$N = \qquad \text{[turns]}$$

Step 9. Calculate the required gap and use iron area A_c found in step 3:

$$l_g = \frac{0.4N^2 A_c \times 10^{-8}}{L} \quad \text{[cm]}$$

$$l_g = \frac{(1.26)(\quad)^2(\quad) \times 10^{-8}}{(\quad)}$$

$$l_g = \qquad \text{[cm]}$$

Gap spacing is usually maintained by inserting kraft paper. However, this paper is available only in mil thicknesses. Since l_g has been determined in cm, it is necessary to convert as follows:

$$\text{cm} \times 393.7 = \text{mils}$$

Substituting values:

$$(\quad)(393.7) = \qquad \text{[mils]}$$

Round off to the nearest even mil and multiply:

$$(\quad)2.54 = \qquad \text{[cm}^2\text{]}$$

Step 10. Calculate the amount of fringing flux. Use the G dimension recorded in step 3:

$$F = 1 + \frac{l_g}{\sqrt{A_c}} \ln\left(\frac{2G}{l_g}\right)$$

$$F = 1 + \frac{(\quad)}{(\quad)} \ln\left(\frac{2(\quad)}{(\quad)}\right)$$

$$F =$$

Step 11. Calculate the new number of turns by inserting the fringing flux:

$$N = \frac{l_g L}{0.4\pi A_c F \times 10^{-8}} \quad \text{[turns]}$$

$$N = \frac{(\quad)(\quad)}{(1.26)(\quad)(\quad) \times 10^{-8}}$$

$$N = \quad \text{[turns]}$$

Step 12. Calculate the winding resistance. Use MLT from step 4 and micro-ohms per centimeter from step 6.

$$R = (\text{MLT})(N)\left(\frac{\mu\Omega}{\text{cm}}\right) \times 10^{-6} \quad [\Omega]$$

$$R = (\quad)(\quad)(\quad) \times 10^{-6}$$

$$R = \quad [\Omega]$$

Step 13. Calculate the copper loss P_{cu}:

$$P_{cu} = I^2 R \quad \text{[W]}$$

$$P_{cu} = (\quad)^2(\quad)$$

$$P_{cu} = \quad \text{[W]}$$

Step 14. Calculate the total ac plus dc flux density:

$$B_m = \frac{0.4\pi N(I_{dc} + \Delta I/2) \times 10^{-4}}{l_g} \quad \text{[T]}$$

$$B_m = \frac{(12.6)(\quad)[(\quad) + (\quad)/2] \times 10^{-4}}{(\quad)}$$

$$B_m = \qquad [\text{T}]$$

Step 15. Calculate the ac flux density:

$$B_m = \frac{0.4\pi N(\Delta I/2) \times 10^{-4}}{l_g} \quad [\text{T}]$$

$$B_m = \frac{1.26(\quad)(\quad) \times 10^{-4}}{(\quad)}$$

$$B_m = \qquad [\text{T}]$$

Step 16. (See design note 9.) Calculate the watts per kilogram for the appropriate core material, and then determine the core loss. Core weight is found in step 3:

$$\text{W/kg} = Kf^{(m)}B_m^{(n)} \quad [\text{T}]$$

$$\text{W/kg} = (\quad)(\quad)^{(\ )}(\quad)^{(\ )}$$

$$\text{W/kg} =$$

$$P_{\text{fe}} = (\text{W/kg})W_{\text{tfe}} \quad [\text{W}]$$

$$P_{\text{fe}} = (\quad)(\quad)$$

$$P_{\text{fe}} = \qquad [\text{W}]$$

Step 17. Calculate the total losses P_Σ:

$$P_\Sigma = P_{\text{cu}} + P_{\text{fe}} \quad [\text{W}]$$

$$P_\Sigma = (\quad) + (\quad)$$

$$P_\Sigma = \qquad [\text{W}]$$

Step 18. Calculate the watts per unit area. The surface area A_t is found in step 3:

$$\psi = \frac{P_\Sigma}{A_t} \quad [\text{W/cm}^2]$$

$$\psi = \frac{(\quad)}{(\quad)}$$

$$\psi = \qquad [\text{W/cm}^2]$$

12.11 TOROIDAL INDUCTOR DESIGN USING THE AREA PRODUCT A_p APPROACH

Design work sheets with the following specifications:

1. Inductance, $L = (\quad)$ [H]
2. dc current, $I_o = (\quad)$ [A]
3. ac current, $\Delta I = (\quad)$ [A]
4. Ripple frequency, $f = (\quad)$ [Hz]
5. Flux density, $B_m = (\quad)$ [T]
6. Core material: _____
7. Core configuration: _____

Step 1. Calculate the energy-handling capability:

$$\text{Energy} = \frac{LI^2}{2} \quad [\text{W-s}]$$

$$I = I_o + \frac{\Delta I}{2} \quad [\text{A}]$$

$$I = (\quad) + \frac{(\quad)}{2}$$

$$I = \qquad \text{[A]}$$

$$\text{Energy} = \frac{(\quad)(\quad)^2}{2}$$

$$\text{Energy} = \qquad \text{[W-s]}$$

Step 2. (See design note 8.)* Calculate the area product A_p:

$$A_p = \left(\frac{2(\text{Energy}) \times 10^4}{B_m K_u K_j}\right)^{(x)} \quad \text{[cm}^4\text{]}$$

Window utilization factor $K_u = 0.4$

$$A_p = \left(\frac{2(\quad) \times 10^4}{(\quad)(0.4)(\quad)}\right)^{(\ )}$$

$$A_p = \qquad \text{[cm}^4\text{]}$$

Step 3. (See design note 4.) Select a comparable area product A_p from the specified core section, and record the appropriate data:

Core _____

$$A_p = \qquad \text{[cm}^4\text{]}$$

$$\text{MLT} = \qquad \text{[cm]}$$

$$A_c = \qquad \text{[cm}^2\text{]}$$

$$W_a = \qquad \text{[cm}^2\text{]}$$

---

*Design notes appear on page 412.

$$A_t = \qquad [\text{cm}^2]$$

$$\text{MPL} = \qquad [\text{cm}]$$

$$W_{\text{tfe}} = \qquad [\text{g}]$$

Step 4. (See design note 8.) Calculate the current density J:

$$J = K_j A_p^{(y)} \quad [\text{A/cm}^2]$$

$$J = (\quad)(\quad)^{(\ )}$$

$$J = \qquad [\text{A/cm}^2]$$

Step 5. Calculate the bare wire size $A_{w(B)}$:

$$A_{w(B)} = \frac{I_o + \Delta I/2}{J} \quad [\text{cm}^2]$$

$$A_{w(B)} = \frac{(\quad) + (\ /2)}{(\quad)}$$

$$A_{w(B)} = \qquad [\text{cm}^2]$$

Step 6. Select a wire size from the wire table (Table 6.1), column 2. If the area is not within 10%, take the next smallest size. Also record micro-ohms per centimeter from column 4.

$$\text{AWG No. \_\_\_\_}$$

$$\text{Bare, } A_{w(B)} = \qquad [\text{cm}^2]$$

$$\text{Insulated, } A_w = \qquad [\text{cm}^2]$$

$$\mu\Omega/\text{cm} =$$

Step 7. Calculate the effective window area $W_{a(\text{eff})}$. Use the window area W_a found in step 3:

$$W_{a(\text{eff})} = W_a S_3 \quad [\text{cm}^2]$$

A typical value for S_3 is 0.75, as shown in Chapter 6.

$$W_{a(\text{eff})} = (\quad)(0.75)$$

$$W_{a(\text{eff})} = \quad [\text{cm}^2]$$

Step 8. Calculate N, the number of turns. Use wire area A_w found in step 6:

$$N = \frac{W_{a(\text{eff})} S_2}{A_w} \quad [\text{turns}]$$

A typical value for S_2 is 0.6, as shown in Chapter 6.

$$N = \frac{(\quad)(0.6)}{(\quad)}$$

$$N = \quad [\text{turns}]$$

Step 9. Calculate the permeability of the core required:

$$\mu_r = \frac{L \times \text{MPL} \times 10^8}{0.4\pi N^2 A_c}$$

$$\mu_r = \frac{(\quad)(\quad) \times 10^8}{(1.26)(\quad)^2(\quad)}$$

$$\mu_r =$$

Now choose a core from the group with a closer permeability.

Core No. _____

$$\mu =$$

$$MH/1000 =$$

Step 10.   Calculate N, the number of turns required:

$$N = 1000 \sqrt{\frac{L}{L_{1000}}} \quad [\text{turns}]$$

$$N = 1000 \sqrt{\frac{(\quad)}{(\quad)}}$$

$$N = \qquad [\text{turns}]$$

Step 11.   Calculate the winding resistance. Use MLT from step 3 and micro-ohms per centimeter from step 6.

$$R = (\text{MLT})(N)\left(\frac{\mu\Omega}{\text{cm}}\right) \times 10^{-6} \quad [\Omega]$$

$$R = (\quad)(\quad)(\quad) \times 10^{-6}$$

$$R = \qquad [\Omega]$$

Step 12.   Calculate the copper loss P_{cu}:

$$P_{cu} = I^2 R \quad [\text{W}]$$

$$P_{cu} = (\quad)^2(\quad)$$

$$P_{cu} = \qquad [\text{W}]$$

Step 13. Calculate the ac flux density:

$$B_{ac} = \frac{0.4\pi N(\Delta I/2)\mu \times 10^{-4}}{\text{MPL}} \quad [\text{T}]$$

$$B_{ac} = \frac{(1.26)(\quad)(\quad/2)(\quad) \times 10^{-4}}{(\quad)}$$

$$B_{ac} = \qquad [\text{T}]$$

Step 14. (See design note 9.) Calculate the watts per kilogram for the appropriate core material, and then determine the core loss. Core weight is found in step 3:

$$\text{W/kg} = K f^{(m)} B_m^{(n)}$$

$$\text{mW/g} = (\quad)(\quad)^{(\quad)}(\quad)^{(\quad)}$$

$$\text{mW/g} =$$

$$P_{fe} = (\text{mW/g})(W_{tfe}) \times 10^{-3} \quad [\text{W}]$$

$$P_{fe} = (\quad)(\quad) \times 10^{-3}$$

$$P_{fe} = \qquad [\text{W}]$$

Step 15. Calculate the total losses P_{Σ}:

$$P_{\Sigma} = P_{cu} + P_{fe} \quad [\text{W}]$$

$$P_{\Sigma} = (\quad) + (\quad)$$

$$P_{\Sigma} = \qquad [\text{W}]$$

Step 16.   Calculate the watts per unit area. The surface area A_t is found in step 4:

$$\psi = \frac{P_{cu}}{A_t} \quad [\text{W/cm}^2]$$

$$\psi = \frac{(\quad\quad)}{(\quad\quad)}$$

$$\psi = \quad\quad\quad [\text{W/cm}^2]$$

Step 17.   (See design note 7.) Calculate the dc magnetizing force (oersteds):

$$H = \frac{0.4\pi NI}{\text{MPL}} \quad [\text{Oe}]$$

$$H = \frac{(1.26)(\quad)(\quad)}{(\quad)}$$

$$H = \quad\quad\quad [\text{Oe}]$$

$$NI = (H)(0.8) \quad [\text{amp-turns}]$$

$$NI = (\quad\quad)(0.8)$$

$$NI = \quad\quad\quad [\text{amp-turns}]$$

DESIGN NOTES

1. When calculating the output power P_o of a transformer, the diodes must be included.
2. When calculating apparent power of a transformer, the primary configuration and each of the secondaries must be taken into consideration (see Chapter 3).
3. The wave-form coefficient K_f is 4.0 for a square wave and 4.44 for a sine wave.
4. The data used in calculating transformers and inductors must include the stacking factor S in order to get the effective value. If the cross section of the core A_c is given in gross terms, then it must be multiplied by the stacking factor S.

$$K_g \times S^2, \quad A_p \times S \quad A_c \times S \quad W_{tfe} \times S$$

5. In a center tap configuration, I_p and/or I_o must be multiplied by 0.707.
6. In the core geometry K_g approach to the design of input inductors or output inductors, the output power P_o refers to the rms current in the inductor times the voltage to ground.
7. See Chapter 5, Table 5.1, for core saturation.
8. See Chapter 3, Table 3.1 for core coefficients x and y.
9. See Chapter 3 for core loss coefficient $f^{(m)}B^{(n)}$.

Index